49 MORE EASY-TO-BUILD ELECTRONICS PROJECTS

Other TAB books by the authors:

No. 1337 *49 Easy-To-Build Electronic Projects*
No. 1339 *101 Easy Test Instrument Projects*

49 MORE
EASY-TO-BUILD
ELECTRONICS
PROJECTS

BY ROBERT M. BROWN & TOM KNEITEL

TAB TAB BOOKS Inc.
BLUE RIDGE SUMMIT, PA. 17214

FIRST EDITION

FIRST PRINTING

Copyright © 1981 by TAB BOOKS Inc.

Printed in the United States of America

Library of Congress Cataloging in Publication Data

Brown, Robert Michael, 1943-
    49 more easy-to-build electronics projects.
    Includes index.
    1. Electronics—Amateurs' manuals.   I. Kneitel, Tom.
II. Title.   III. Title: Forty-nine easy-to-build electronics projects.
TK9965.B742   621.381   81-9205
ISBN 0-8306-0021-3   AACR2
ISBN 0-8306-1347-1 (pbk.)

Preface

This is a book of easy and entertaining electronic projects. Forty-nine projects are included in this volume, most of which you can build in one evening and for less than $10. The projects range from fun types, such as an electronic rifle range; through practical types, such as a fence charger; to amusement types, such as a transistor tickler. Other devices, particularly the perpetual beeper and the free-power AM radio receiver, are suitable as school science projects.

The parts can be purchased for low cost and many can be obtained by salvaging old radios and television receivers. Generally, parts substitutions are feasible but the substitutes should be kept as close as possible to the specified values. Voltage ratings of the electrolytic capacitors are given in the parts lists.

If you are new to electronics projects building, we suggest that you first read the following "Tips to the Beginner," which will acquaint you with the basics of the subject. A substitution guide, resistor and capacitor color codes and list of electronics symbols are included in the appendices.

Although we enjoyed constructing the projects as they are discussed here, you may think of possible modifications. We heartily recommend that you go ahead and modify the circuits to your own personal ends.

Note: The electronics parts numbers listed in this book are intended only as a guide in selecting the correct types of electronic components.

The reader is referred to the many electronic substitution guides, such as TAB's two volumes, Nos. 1470 and 1471.

Robert M. Brown
Tom Kneitel

Table Of Contents

Tips For The Beginner

If you are a beginner in electronics projects building, you would be well advised to start with the simplest projects, i.e., those using neither tubes nor transistors. Recommended circuits are those relying on, say, resistors, capacitors, batteries, and neon bulbs.

First, you will need a good supply of hookup wire. For circuits the power to which is supplied by 9-volt and 22½-volt battery packs you can use any insulated hookup wire available. For higher-power circuits in this book such as those supplied by 120 VAC, insulated line cord pulled apart is ideal. Always keep hookup wires as short as possible, yet never so short as to create a danger of short-circuiting should the device be jarred or dropped.

A pair of long-nosed pliers, a pair of conventional pliers, a small screwdriver, a hand or power drill, and an assortment of wrenches for installing nuts and bolts are highly useful.

Resistor values are indicated in ohms on the schematics and are ½ watt unless otherwise specified in the parts lists. If you are slow on determining resistor values from resistor markings, consult the resistor color-code chart in Appendix B. If you do not have a resistor of a certain required value, you may be able to series or parallel connect other-value resistors (while observing power requirements) to achieve the required value. Also, don't feel hamstrung because you can't find, say, a 25K potentiometer; chances are that a 35K or 50K potentiometer will work just as well. Similar approximations to fixed-value resistors are also possible, and a rule of thumb for resistor substitutions is always to exceed the specified resistance; never go under it. In any case, keep the substitutes close to the original values.

Although high-capacitance capacitors are usually typographically marked, most low-capacitance types are color coded. The capacitor color codes are given in Appendix C. The "μF" types used are usually tubular and their values are printed on the sides of the capacitors. Electrolytic capacitors, which are used frequently, also have a voltage rating, which is given in the parts lists but not on the schematic diagrams. The plus sign on the electrolytic capacitor symbol indicates the hot side of the capacitor. This side is also indicated on the capacitor itself. Electrolytic capacitors deteriorate with age, so be alert to this possibility of malfunction.

Batteries are also often used in the projects, since they supply smooth DC and permit portability of the device. You should not attempt, however, to power a tube-type device that makes use of relays and calls for 6 VDC with four 1½-volt microminiature hearing-aid cells. Logic will tell you that a much larger battery pack will be required. Most transistor circuits, however, will perform very well on small batteries normally used for powering semiconductor devices.

Most of the projects are made of ordinary components which can be salvaged from worn-out household electronic gear. Thus, for maximum economy, you may cannibalize radios, TV receivers, hi-fi equipment, etc. when they are to be disposed of. Remove all components—even tube sockets and switches.

A substitution guide for transistors, tubes, and diodes, used in this book is given in Appendix A. Standard schematic symbols are given together with their meanings in Appendix D.

Heat is the enemy of all electronic components, be they capacitors, tubes, or whatever. If you intend to do a lot of soldering, it would be wise for you to purchase a light-weight soldering gun or small soldering iron or pencil. You will need a soldering tip which is long and narrow and which does not get hotter than is necessary to melt the solder. When you get the more complex circuits using semiconductors, you will need to use a heat sink to prevent heat from ruining the semiconductors.

Keep a good supply of cigar boxes handy which can be filled with machine screws and nuts suitable for use in mounting components and terminal strips. As an added feature, you will find many projects that you can build right *into* the cigar boxes!

Cigarette-Pack Flasher

One of the most fascinating things in electronics that a person can experiment with is the neon bulb. Generally available for about 25 cents each (or less), the tiny NE-2 bulb is an interesting item. All that is necessary is a handful of capacitors and resistors, and you will be blinking and winking in no time.

What few people realize about neon flashers is that, although they generally require 90 VDC or more to "Fire," they draw so little current that large, heavy-duty 90-volt B cells *do not* have to be installed to get them to operate. A glance at Fig. 1 and Table 1 will illustrate this. In this instance, four 22½-VDC hearing-aid batteries are wired in series.

This amazing gadget, requires only three resistors, three capacitors, three neon bulbs, and the four-hearing-aid cells just described. The "chassis" will be an empty cigarette pack, something, perhaps, on the order of the hardpacks. This will give you a bit of near-cardboard support on which to mount your bulbs and a durable case for housing circuitry and batteries.

Note that there is no on-off switch—it is unnecessary. All three bulbs will flash alternately and in unusual patterns for periods of up to one *year* without requiring fresh battery replacement! (Incidentally, this gadget is a great toy for children.)

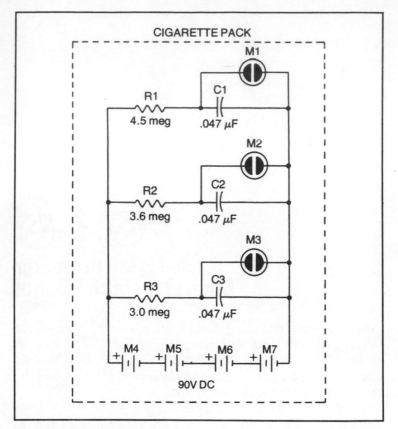

Fig. 1. Cigarette-pack flasher.

Table 1. Parts List for Cigarette-Pack Flasher.

Item No.	Description
C1, C2, C3	0.47-μF capacitors.
M1, M2, M3	NE-2 bulbs.
M4, M5 M6, M7	22½-VDC batteries.
R1	4.3-megohm resistor.
R2	3.6-megohm resistor.
R3	3.0-megohm resistor.

2

"Black-Light" Projector

Here is a gadget that can be assembled in a couple of hours, yet will provide endless hours of use on the job. If you are not familiar with ultraviolet or "black" light, you are truly missing out on a fascinating scientific phenomenon. Commercially, black lights are used for chemical analysis, prospecting, and even criminal investigation. For our purposes, however, the amazing black light will reveal curious colors and glows of ordinary household objects which you would never suspect otherwise. For example, fluorescent lamps, dyes, inks, and even popular laundry detergents, take on a whole new appearance under the sole scrutiny of a black light. Any object with any degree of characteristic fluorescence will take on a weird coloration and actually glow in the dark. Certain rocks and minerals will also emit strange light colorations when you put them near your black-light projector.

The circuit is uniquely simple and straightforward. The parts are available through several commercial outlets, and no intermediary transformer or the like is required. See Fig. 2 and Table 2.

Curious experimenters might be interested to know the fact that ultraviolet radiation falls below X-rays and just above visible light on the electromagnetic spectrum.

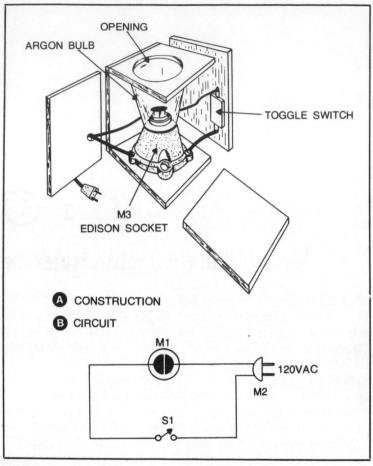

Fig. 2. Black-light projector.

Table 2. Parts List for Black-Light Projector.

Item No.	Description
M1	2-watt argon bulb.
M2	AC wall plug and clamp.
M3	Edison socket.
S1	Spst switch.
	Plywood or other suitable material for housing bulb and socket.

3

Unusual Musical Effects Device

If you have any motor-driven music boxes around, you can construct an unusual modern electronic music box which can be played through any standard audio amplifier. Its chimelike quality is ideal for outdoor reproduction, and there will never be a need to flip the records over or change the tapes. But best of all, you will be producing a unique sound.

It is suggested you build your musical effects device in a wooden box (for resonance), although you can utilize a standard minibox or other metallic housing. Conventional (inexpensive) contact microphones are coupled mechanically to the tone plate of each of the three music boxes. The inexpensive motors tend to keep the boxes playing and provide the normal wind-up action.

The construction is quite simple, provided you stick to the outline shown in Fig. 3. To operate the effects box, connect the output of this device to the mike input of your hi-fi or other audio amplifier. Turn one of the music boxes on, and adjust the motor speed by attaining the desired setting on the potentiometer. Then set the volume you want on the amplifier. Whenever you wish, switch that music box off, and turn on another. See Table 3.

Incidentally, you can obtain interesting results if you turn two boxes on simultaneously. This sometimes simulates a "round," but only if you can synchronize both movements at precisely the right moment.

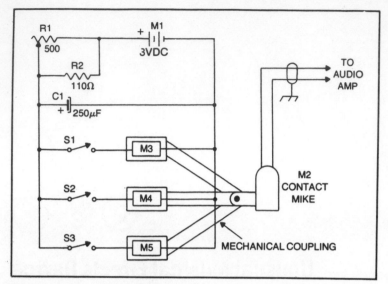

Fig. 3. Unusual musical effects device.

Table 3. Parts List for Unusual Musical Effects Device.

Item No.	Description
C1	250-μF, 15-w VDC electrolytic capacitor.
M1	3-VDC battery.
M2	Guitar or other contact-type microphone.
M3, M4, M5	Motor-driven music box movements.
R1	500-ohm, 3-watt potentiometer.
R2	110-ohm resistor.
S1, S2, S3	Spst switches.
	Three metal contact clips.
	One microphone connector.
	Wooden box or minibox.

Tilter Box Oscillator

Here is a gadget you cannot buy anywhere, yet it serves a useful and entertaining purpose: that of keeping you awake when driving, working, or whatever. Best of all, the curious will never know more than the fact that you have a common hearing-aid type earpiece in your ear.

The idea is quite simple. Taped to the rear of the earpiece is a subminiature mercury switch, positioned in such a way as to *not* make contact when worn normally. The minute, though, that you begin to slump over from weariness, *beep*! This awakens you and you resume a normal driving position, at which time the tone blast ceases.

You can build the beep box into any handy plastic container— the kind that small machine screws come in is ideal for this purpose. For the transformer, you can use a Lafayette 99 H 6126 or similar type. The battery can be a penlite cell or flashlight battery, depending upon what size of container you are using.

Incidentally, wire the switch and earphone exactly as shown. This will allow you to use the normal flexible earpiece cabling that comes with most transistor-radio and hearing-aid phones, an aid when you are wearing the tilter. Stiffer wire, of course, will work, but will prove annoying after a while. See Fig. 4 and Table 4.

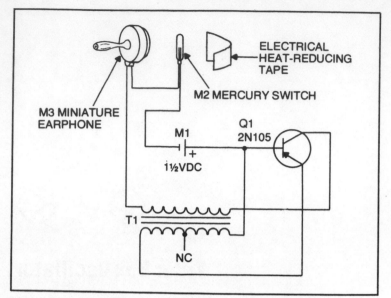

ELECTRICAL HEAT-REDUCING TAPE

M2 MERCURY SWITCH

M3 MINIATURE EARPHONE

M1

Q1 2N105

1½VDC

T1

NC

Fig. 4. Tilter box oscillator.

Table 4. Parts List for Tilter Box Oscillator.

Item No.	Description
M1	1½-VDC cell.
M2	One subminiature mercury switch.
M3	One miniature transistor-radio earphone, 2K-3K impedance.
Q1	2N105 transistor.
T1	Transistor transformer. (Lafayette 99 H 6126 or equivalent.)
	One strip of heat-reducing or other electrical-type tape to secure earpiece to switch.

5

Radio-Band Metronome

Here is a project guaranteed to delight builders, who have various ideas on what they want from it. Basically, the radio metronome is a tiny transistor transmitter which, instead of sending audio or modulation to a remote receiver, sends loud "tick-tock" sounds.

This device costs much less to build than a commercial music-store-variety metronome you could buy. Also, you can *vary* the frequency of the "tick-tocks." Another possible use is as a device to help get you to sleep.

It can be built into any kind of housing which you select to use, and merely placed in close proximity to your AM table radio. Pick out a clear spot on the dial (someplace between two local stations, a spot not occupied) and simply adjust L1 until you hear your radio-band metronome's signal. Once the frequency has been set, adjust potentiometer R1 for the frequency of "tick-tocks" you desire.

Should more distance be desired, run a short length of antenna wire from the junction of Q1-C1-L1 indicated by the arrow in Fig. 5. You may have to readjust L1 after doing this, but you will have a much stronger signal. Also, see Table 5.

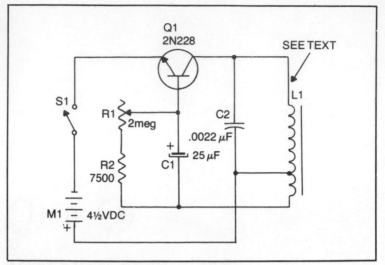

Fig. 5. Radio-band metronome.

Table 5. Parts List for Radio-Band Metronome.

Item No.	Description
C1	25μF, 15-w VDC electrolytic capacitor.
C2	.0022-μF capacitor.
L1	700 turns No. 25 enameled wire wound evenly on ¼-inch iron slug 1½ inches long. Tapped at 500 turns (this winding group is soldered across C2).
M1	4½-VDC battery.
Q1	2N228 transistor.
R1	2-megohm potentiometer.
R2	7.5K resistor.
S1	Spst switch.

6

Winking Night Light

You can make a useful night light for the baby's room for the cost of a single .0047-μF capacitor, and NE-2 neon bulb, an 11-meg resistor and a pair of 45-VDC batteries. If you can come up with another way of supplying 90 VDC, you can make this device for practically nothing.

The interesting thing about this device is that it will virtually run forever. The NE-2 bulb consumes so little electrical current that between "flashes" the batteries tend to recharge themselves. One such device constructed by the authors over two years ago is still flashing today, although the flash rate has dropped somewhat. To compensate for this without replacing the batteries merely change the RC circuit by altering the value of R1. See Fig. 6 and Table 6.

You can build this device into just about any container you wish, although we remind you that anything in a youngster's room is apt to take quite a beating. For this reason, it is suggested you buy a toy and modify it for your own purposes; the selection of the toy will pretty much depend on how large your batteries are and how destructive your youngster is. At any event, once constructed, the night light should be sealed shut (a bit of heat does wonders with plastic).

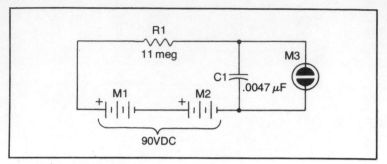

Fig. 6. Winking night light.

Table 6. Parts List for Winking Night Light.

Item No.	Description
C1	.0047-μF capacitor.
M1, M2	45-VDC batteries.
M3	NE-2 bulb.
R1	11-megohm resistor.

7

Electronic Chirper

Here is a project that may require a bit of expenditure on your part for components, but you will end up with a device that really puts out some weird sounds. The "chirper" imitates the electronic sirens that are so popular in big cities today. It has a control that can be used to change the pitch of the basic output from low frequencies to high.

If the speaker (4 to 6 inch should be adequate) is mounted in the same box with the circuitry, it will determine the overall size of the unit. The other components are small—they can be mounted on a small board and tucked away in corner somewhere. Do allow for the batteries, however. They will take up more room than the transistors, capacitors, and resistors put together.

Basically the circuit is a free-running multivibrator. Control R4 varies the frequency of oscillation from about 1000 hertz to 10,000 hertz. The chirping effect is accomplished by a low-frequency oscillator using an NE-2 bulb (M3). This signal provides base bias for transistor Q1. If capacitor C1 is shorted, the chirping is eliminated, and the circuit becomes a simple variable-frequency multivibrator. Transistor Q3 is the driver stage, and Q4 is the output transistor. See Fig. 7 and Table 7.

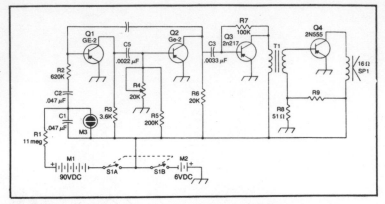

Fig. 7. Electronic chirper.

Table 7. Parts List for Electronic Chirper.

Item No.	Description
C1, C2	.047-μF capacitors.
C3	.0033-μF capacitor.
C4	.0022-μF capacitor.
M1	90-VDC battery.
M2	6-VDC battery.
M3	NE-2 bulb.
Q1, Q2	GE-2 transistors.
Q3	2N217 transistor.
Q4	2N555 transistor.
R1	11-megohm resistor.
R2	620K resistor.
R3	3.6K resistor.
R4	20K potentiometer.
R5	200K resistor.
R6	20K resistor.
R7	100K resistor.
R8	51-ohm resistor.
R9	3K resistor.
S1 (A, B)	Dpst switch.
SP1	16-ohm speaker.
T1	Transistor interstage transformer: 100-ohm primary, 10-ohm secondary. (Stancor TA-2 or equiv.)

8

Electronic Rifle Range

Here is a project that will provide fun for the whole family. It uses a flashlight-equipped toy rifle and phototube target. A buzzer alarm is used to signify that the bull's-eye has been hit. The project is a game that can be built almost entirely from available parts, runs off household 120 VAC, and employs no semiconductors.

Operating on a light-sensitive principle, the 927 tube is positioned directly behind the target. The bull's-eye is hollow, permitting light to strike the phototube (V1). This tube in turn feeds a 6BM5 tube which acts to activate the relay (K1) that controls current reaching the alarm bell or buzzer. The bell will sound only when light is directly on the bull's-eye.

The circuit, while extremely simple to build, does use 120 volts, so adequate precautions should be taken. The 120-volt line is at no time grounded. Rubber grommets are employed to prevent the insulation wearing and consequently shorting against the chassis.

The "firing" takes place from a flashlight that has been mounted on top of a toy rifle. A small contact switch can be positioned near the rifle trigger with wires running to the flashlight, so that it is possible to turn the flashlight on at the flick of a switch. A push-button miniature spst switch is ideal for this purpose. See Fig. 8 and Table 8.

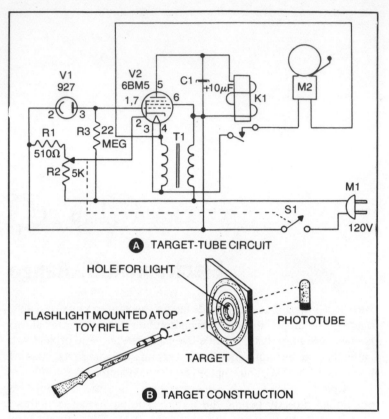

Fig. 8. Electronic rifle range.

Table 8. Parts List for Electronic Rifle Range.

Item No.	Description
C1	10-μF, 150-2 VDC electrolytic capacitor.
K1	4K relay.
M1	AC wall plug, cap, and clamp.
M2	6-VAC buzzer or alarm.
R1	510-ohm, 2-watt resistor.
R2-S1	5K potentiometer with spst switch.
R3	22-megohm, 2-watt resistor.
S1	Spst switch on R2.
T1	6-volt filament transformer. (Triad F-14X or equiv.)
V1	927 phototube.
V2	6BM5 tube.
	Two standard 7-pin miniature tube sockets.

26

9

Electronic Fish Lure

This is the original electronic fish lure, not those quieted-down commercial versions advertised in sporting and fishing magazines. This gadget is the one that started the rage and the fish have not caught on yet.

The reason this lure is so much more effective than the rest is that it contains a foolproof method of attracting fish: it not only imitates the sounds of wet bugs milling about at the surface, but it also flashes annoyingly in a fashion that just cannot be ignored. Intermittent tiny flashes of light penetrate for hundreds of feet, drawing the curious by sight as well as by sound. The latter feature, the buzzing sound, is peculiarly resonant, making it all the more loud and annoying to fish.

You must build this one into a clean, empty peanut-butter jar with the buzzer circuitry mounted upside-down on the cover of the jar. Be careful not to drill in such a way as to allow water to leak in through the cap. The blinker circuit is simply laid on the bottom of the jar and arranged so that the bulbs are clearly visible from under the water.

To operate, merely unscrew the cap, throw S1 (a dpst switch) on, put the cover back on, and drop the unit in the water in the vicinity of where you will be fishing. Allow 15 minutes, then start reeling them in! See Fig. 9 and Table 9.

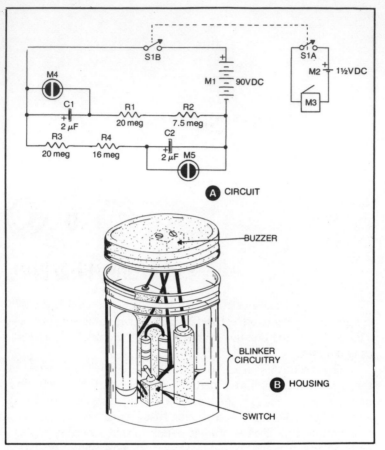

Fig. 9. Electronic fish lure.

Table 9. Parts List for Electronic Fish Lure.

Item No.	Description
C1, C2	2-μF, 150-w VDC electrolytic capacitors.
M1	90-VDC battery.
M2	1½-VDC cell.
M3	1½-volt high-frequency buzzer.
M4, M5	NE-2 bulbs.
R1, R3	20-megohm resistors.
R2	7.5-megohm resistor.
R4	16-megohm resistor.
S1 (A,B)	Dpst switch.
	One peanut-butter jar.

Toy Telephone Amplifier

The only real problem with the working-type toy telephones normally found in toyshops and department stores is that the volume is much too low. Build this simple modification, however, and you will have plenty of volume.

Follow closely the "original" and "additional" sections in Fig. 10. Compare the wiring and parts placement in the "original" telephone with your own toy telephone circuitry. Mark clearly in the book the proper color codes to coincide with your own system, so that you can identify the wires. (Mark these wire colors on both "original" and "additional" diagrams.) Also note any possible differences between this circuit and your own. Once this has been established, you are ready to proceed. See Table 10.

Next, solder the 2N331 transistor, Q1, and resistor R1 in place, taping the joints and other areas of exposed lead lengths so as to prevent any shorting. If your toy telephone is plastic, take precaution with the soldering-iron heat so that you do not melt anything! Make certain you use the proper battery polarity; if you reverse it, you will ruin the transistor.

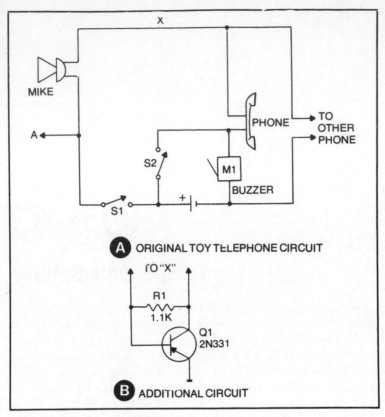

Fig. 10. Toy telephone amplifier.

Table 10. Parts List for Toy Telephone Circuit.

Item No.	Description
Q1	2N331 transistor.
R1	1.1K resistor.

Coin Battery

In addition to drawing interest from a bank, you can draw the interest (and amazement) of your friends by demonstrating how you can power an electronic circuit with nothing but an ordinary coin! You can "convert" a 50-cent piece into a working battery.

Of course you will not produce a lot of power, but you can get enough to trigger a junction transistor into oscillation, allowing you to key the thing for a reliable code practice oscillator. Even if you could not care less about International Morse Code, your friends' faces will show some display of shock when they hear the beeps you produce when depressing the key—particularly in full view of the 50-cent piece.

The best way to construct your "battery holder" is to use bare copper wire bent as shown in Fig. 11. Then you merely slip the coin between the "clips" and produce—instant current.

The secret is to cut a disk of ordinary paper and soak it in a saltwater solution until it is just about saturated. Then cover one side of the coin (either heads or tails) with the disk. The salt-paper side will be your negative (−) terminal, and the clear side of the coin the positive (+) terminal. Also, see Table 11.

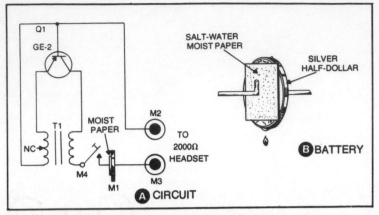

Fig. 11. Coin battery.

Table 11. Parts List for Coin Battery.

Item No.	Description
M1	Half-dollar coin.
M2, M3	Test jacks.
M4	Code practice key.
Q1	GE-2 transistor.
T1	Transistor driver transformer: 20K primary, 2K secondary. (Argonne AR-103 or equiv.)

32

Electronic Metal Locator

This project requires a few more parts than most projects in this book, but what you will wind up with is an amazingly effective metal locator you can use for finding wall studs, underground pipelines, and even buried treasure. Best of all, if you have a well-stocked spare-parts bin you can build it for little more than the price of the metal box that houses the circuitry. This is inexpensive, because unlike most metal locators, this one has a standard table-radio antenna coil backplate for the critical metal-locating sensor. Therefore, the major part of your total outlay for parts is nothing more than a quick look into the parts bin. You will use the entire backplate, feeding the antenna coil wires through an empty paper-towel tube to the locator box.

To achieve the desired degree of sensitivity, you must "edge" the antenna coil with a strip of normal household screening. See Fig. 12 and Table 12. This is simple, since just about every time you pull a loose wire on a window screen, you not only get the ripped wire, but also a few inches of side "hairs" over the circular coils in a more or less protective manner. Cut off the excess hair length. You only want to cover the coils. Then scotch-tape the whole arrangement so as to hold it in place. Do not make electrical connection an any point. The screen rib and hairs should not make connection in any electrical way with the antenna coil or the locator circuit.

Once the primary circuit has been installed in the box and the tube-antenna-coil unit mounted as shown, it is time to test your locator. After tuning S1 on, increase the R5 setting to maximum sensitivity. Bring the locator physically close to a large metallic object for testing purposes. Now plug in the earphones. If you hear a buzzing or tone in the phones, back off the locator from the metal object. The tone should trail off until you do not hear it again.

If you get no sound at all, or if the sound is weak, adjust the setting of L1 (by adjusting the loopstick ferrite core) slightly until maximum signal is heard. Then tune C6 also for maximum signal. Now you should be ready for action.

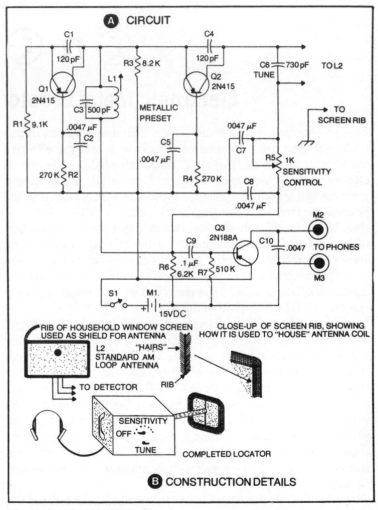

Fig. 12. Electronic metal locator.

Table 12. Parts List for Electronic Metal Locator.

Item No.	Description
C1, C4	120-pF capacitors.
C2, C5, C7, C8, C10	.0047-μF capacitors.
C3	500-pF capacitor.
C6	730-pF variable capacitor. (Two gang 365-pF sections in parallel.)
C9	1-μF capacitor.
L1	Ferrite-rod antenna. (Lafayette 32 C 4 106 loopstick or equiv.)
L2	Standard AM antenna.
M1	15- VDC battery.
M2, M3	Test jacks.
Q1, Q2	2N415 transistors.
Q3	2N188A transistor.
R1	9.1K resistor.
R2, R4	270K resistors.
R3	8.2K resistor.
R5	1K variable resistor.
R6	6.2K resistor.
R7	510K resistor.
S1	Spst switch.

Stand a foot or so away from a metallic object and repeat this procedure. You will probably find some fine-tune adjustments have to be made with L1 and C6.

In normal use, you will adjust R5 for an adequate sensitivity level (you will not want it wide open all the time) and tune with C6.

 13

Perpetual Beeper
and Meter Deflector

Here is a truly amazing gadget you can put together for very little cash outlay, and which will practically run forever. Best of all, it is portable and can be arranged interestingly on a display board for demonstration purposes.

An inexpensive 2N105 junction transistor is used, and, once power is applied to the circuit, a continuous string of intermittent "beeps" will emit from a pair of earphones. If you like, you can add a one- or two-transistor follower stage for speaker operation, although this will increase the amount of current drain on the battery and force you to discard the perpetual feature of the circuit. In any case, you will get "beeps" plus an interesting "backwards-meter" effect. The meter will read full scale, *deflecting* downward occasionally rather than rising.

In any case, you can adjust the rate of meter deflections and tone pulses by simply setting R1 to the point desired for maximum effect.

Note that the circuit will not work unless the 1N38B diode is connected exactly as shown with regard to cathode-anode polarity. See Fig. 13 and Table 13.

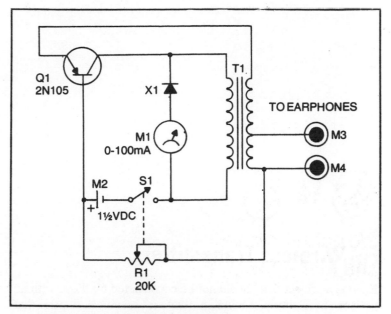

Fig. 13. Perpetual beeper and meter deflector.

Table 13. Parts List for Perpetual Beeper and Meter Deflector.

Item No.	Description
M1	0-100-mA milliammeter.
M2	1½-VDC cell.
M3, M4	Test jacks.
Q1	2N105 transistor.
R1-S1	Switch control: 20K variable resistor ganged with spst switch. (IRC Q11-119 or equiv.)
T1	Transistor drive CT transformer. (Argonne AR-103 or equiv.)
X1	1N38B diode.

FM Wireless Transmitter

Here is a project that should not be constructed by those with no patience for electronics building. While it basically uses parts that can be obtained from any nearby radio-TV store (you can strip a non-working radio or TV set for parts), it is a bit more of a project to attempt and consequently presents more room for error than the others in this book. For this reason building this project takes a great deal of patience. The payoff is a sensationally powerful FM transmitter that will allow you to plunk a hefty signal down anywhere on the 88-108-MHz band with "broadcast" quality audio. Additionally, many features are included that can either be utilized, or simply left off your model, depending on how fancy you want your wireless transmitter to be.

Typical of these additions is the VU meter, shown as M1 in Fig. 14. Should you wish not to include this, merely omit both M1 and resistor R14 entirely. If you want to "monitor" your own audio, you can hook up a pair of earphones between ground and the junction of the shielded cable, C5, and R14, and insert a 5K potentiometer across the monitor lines. In this fashion you can determine how far to advance volume control R11 without introducing audio distortion. By the same token, you can use a 1N38B germanium diode across your earphones to permit monitoring "wireless" transmissions. In this way, you are receiving the actual on-the-air signals from the transmitter by bringing the diode (and coil, if you want to be thorough) close to the broadcasting antenna. In many respects, this method of monitoring is prefera-

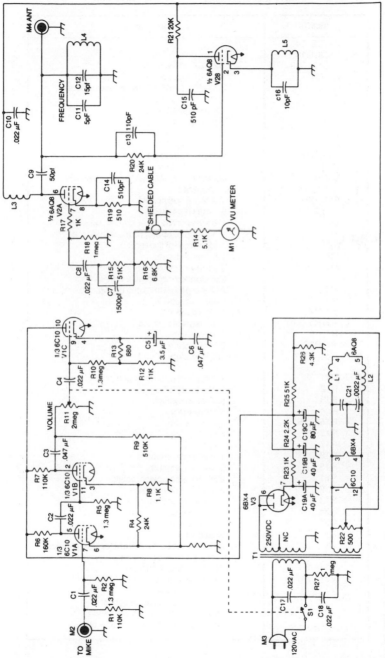

Fig. 14. FM wireless transmitter.

39

Table 14. Parts List for FM Wireless Transmitter.

Item No.	Description
C1, C2, C4, C8, C10, C17, C18	.022-μF capacitors.
C3, C6	.047-μF capacitors.
C5	3.5-μF, 75-wVDC electrolytic capacitor.
C7	1500-pF capacitor.
C9	50-pF capacitor.
C11	5-pF capacitor.
C12	15-pF variable capacitor.
C13	110-pF capacitor.
C14, C15	510-pF capacitors.
C16	10-pF capacitor.
C19 (A, B, C)	40/40/80-μF, 450-wVDC electrolytic capacitor.
C20, C21	.0022-μF capacitors.
L1, L2, L3	3-3-mH rf chokes.
L4	5½ turns of No. 16 enameled wire on ¼-inch diameter form, spaced the diameter of the wire.
L5	1.5-mH rf choke.
M1	VU meter.
M2	Microphone receptacle with locknut.
M3	AC wall plug with cable clamp.
M4	Coax receptacle.
R1, R7	110K resistors.
R2, R5, R10	1.3-megohm resistors.
R3, R8	1.1K resistors.
R4, R20	24K resistors.
R6	160K resistor.
R9	510K resistor.
R11	2-megohm potentiometer.
R12	11K resistor.
R13	680-ohm resistor.
R14	5.1K resistor.
R15, R25	51K resistors.
R16	6.8K resistor.
R17	1K resistor.
R18, R27	1-megohm resistors.
R19	510-ohm resistor.
R21	20K, 10-watt power resistor.
R22	500-ohm potentiometer.
R23	1K, 1-watt resistor.
R24	2.2K resistor.
R26	4.3K resistor.
S1	Spst switch.
T1	120:250-VAC power transformer. (Stancor PS-8416 or equivalent.)
V1 (A,B,C)	6C10 compactron tube.
V2 (A,B)	6AQ8 twin-triode tube.
V3	6BX4 duodiode.
	One compactron tube socket.
	One 7-pin miniature tube socket.
	One 9-pin miniature tube socket.

ble, since you are hearing what the listener is hearing *after* the modulation has passed through the 6AQ8 oscillator.

By the way, the junction point of R23 and R24 should be the only point feeding plate voltage to the tube. If, for example, you take your B+ from the cathode of the 6BX4 (pin 7), the voltage would cause your transmitter to possibly violate Part 15 of the FCC Rules and Regulations due to the fact that it would be generating too much RF power output. For the same reason, the antenna should be short. If the plate power increases, or the antenna is longer than just a few feet, your signals will be heard all over town and the FCC will be looking for you.

The only two expensive components (if you do not have them already) are the compactron, a triple-section tube manufactured by GE, and the power transformer.

The operation is simple. Just adjust C12 so that your signal appears on a blank space in the 88–108-MHz band. Stick to the coil values recommended, and you will have no trouble. Also, see Table 14.

 15

Neon-Bulb Wattmeter

Here is a gadget that could easily be called the simplest, most inexpensive wattmeter in the world. Instead of complex circuitry filled with Nixie tubes and semiconductors, it relies entirely on neon bulbs to do the work. When it is completed, you will be able to feed audio into the circuit (that would normally be fed to a speaker), and the correct bulb will light signifying "3 watts" or the like.

Note that the presetting of a potentiometer determines when the neon bulb fires. The best way to do this is simply to feed audio into the circuit of a *known* wattage level. This will mean, of course, that you will be running the amplifier in question wide open—full volume. If you know, for example, that this amplifier puts out a maximum of 4 watts, simply adjust the corresponding potentiometer in such a way that the bulb *just lights* with this input. Next, move to another neon bulb and repeat this procedure with a higher-wattage known source, for example a table radio with an 8-watt audio output.

If it is difficult for you to calibrate your instrument in this manner, you can do it with a variable-wattage source. For purposes of calibration, you can use a 12.6 VAC filament transformer with a 100-ohm potentiometer across it. The primary is connected to the line and a variable voltage may be tapped off the potentiometer. This voltage may be applied to the amplifier terminals according to $E = \sqrt{PZ}$, where Z is the amplifier output impedance and P is the wattage to be fed into the wattmeter for calibration purposes. For example, for an output impedance of 4 ohms and a wattage of 4 watts, the voltage must be $E = \sqrt{4 \times 4} = 4$ volts. Be sure that the wattage does not exceed the universal output transformer rating. See Fig. 15 and Table 15.

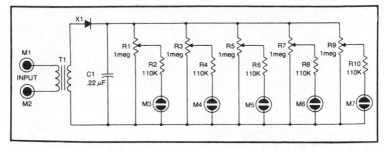

Fig. 15. Neon-bulb wattmeter.

Table 15. Parts List for Neon-Bulb Wattmeter.

Item No.	Description
C1	.22-μF, 450-w VDC capacitor.
M1, M2	Test jacks.
M3, M4, M5, M6, M7	NE-2A neon bulbs.
R1, R3, R5, R7, R9	1-megohm potentiometers.
R2, R4, R6, R8, R10	110K resistors.
T1	Universal output transformer. (Lafayette 33 H 7503 or equiv.)
X1	100-mA, 400-PIV silicon diode. (Lafayette 19H 5001 or equiv.)

43

 16

Flashlight Radio Transmitter

With two gadgets constructed identically from the circuit shown in Fig. 16, you can simply aim one flashlight at the other and virtually talk over the beam. In practice, of course, it's a lot easier to construct your flashlight transmitters in two sections: (1) the main unit with the magnifying lens mounted in a hole in the side of the chassis or box (to admit magnified light so as to "trigger" a 2N280 transistor), and (2) the flashlight itself connected to the transmitter by a pair of wires, yet maintaining mobility for purposes of aiming the beam. Across the area between two houses, for example, you could operate your transmitter from an open upstairs window, aiming your flashlight so as to "hit" the magnifying lens (or bullseye) of the sill-mounted transmitter in a friend's upstairs window. Particularly at night when you aren't troubled with sunlight, you can communicate over considerable distances in this fashion. It is important, however, that your flashlight beam hit the target (magnifying lens) right on the nose.

The secret of this transmitter is the 2N280 garden-variety transistor referred to above. If you are careful, you can scrape the paint off the outside of the transistor, exposing the internal parts of the transistor to all onlookers. If you've done a clean job of it, you now have an inexpensive light-sensitive sensor plus an amplifier, all rolled into one. You need only position the transistor under the magnifier so that the pin-point of light created by your friend's flashlight is directed smack dab onto the 2N280. S1 functions as the transmit-receive switch, and S2 is part of potentiometer R5, acting as your off-on-volume control.

Naturally, the greater the amount of light you beam out, the farther you will be able to communicate. One trick is to rig up a

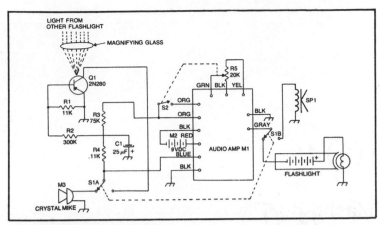

Fig. 16. Flashlight radio transmitter.

Tensor lamp with a hefty battery pack and a flashlight lens for directivity. This provides a much more intense light, although it means more work. You will have to find some way of isolating the power source from the prime supply voltage. If you want to do this, feed the Tensor bulb 12 VDC direct, bypassing the AC transformer in the base of the lamp. Rig up a series of 1½-VDC batteries to arrive at the desired 12 volts, tapping off at 3 volts to feed the circuit in the accompanying diagram. Under no circumstances feed more than 3 volts to the audio amplifier. Also, see Table 16.

Table 16. Parts List for Flashlight Radio Transmitter.

Item No.	Description
C1	25-μF, 25-wVDC electrolytic capacitor.
M1	Audio amplifier module. (Lafayette 99 G 9037.)
M2	9-VDC battery.
Q1	2N280 audio transistor with paint removed from glass case, exposing internal junction to act as phototransistor when light is present under magnified conditions.
R1, R4	11K resistors.
R2	300K resistor.
R3	75K resistor.
R5-S2	20K potentiometer with spst switch.
S1 (A,B)	Dpdt switch.
SP1	8-ohm transistor radio speaker. (Lafayette 99 T 6036 or equiv.)
	One flashlight.
	One magnifying glass.

Solar TV Reception Booster

Here is a versatile rooftop TV reception booster that will add a full 18 dB of gain to the incoming TV signals at your receiver. This means no more ghosts, clearer pictures, and perhaps even reception of a distant station you could not get before. Best of all, this gadget is powered by an inexpensive solar cell that will virtually run forever without requiring replacement. Additionally, the circuit is designed with a charger circuit and reserve pack.

You can build it as is, or only include as much as you would like. For example, if you break the circuit where the X's are indicated, your booster will thrive entirely off its solar battery. If you break it at the two Y's, the solar cell charges a 4½-volt battery pack during the day so the booster will work just as well at night as in direct sunlight. If you include all the circuit (taking into account the indoor AC supply shown), you can occasionally recharge the 4½-volt pack just in case you have had a protracted period of rain or overcast weather which prevented the solar cell from sending its energy down to the reserve battery.

Construction is not complicated, but since you are dealing with both vhf and uhf signals, you will have to keep all leads extremely short and direct. In fact, you would do best to cram the parts close together so as to cut down on the length of interconnecting wires. Looking at the schematic, you will note that Q1 must be grounded; this is no problem; however, since the HEP-3 comes with four leads instead of the regular three. Turning the transistor

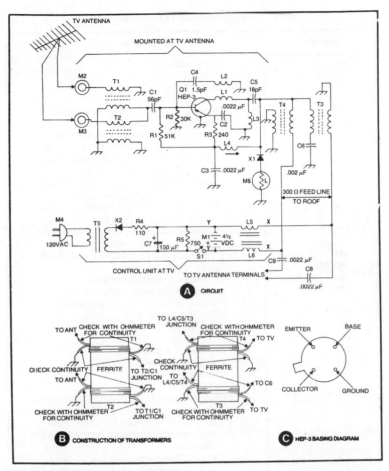

Fig. 17. Solar TV reception booster.

upside down, you will notice an arrangement of leads as shown and identified in Fig. 17.

Take care in constructing your coils that you duplicate exactly those turns and techniques recommended in Table 17.

Once complete, you can test out your booster by hooking it to the TV set and placing a lit 60-watt lamp bulb in rather close proximity to M5. As the lamp is turned on, you should get a much-improved picture. Adjust the best reception on all channels by tuning L4 and the coupling between L1 and L2. Now mount at the TV antenna, positioning M5 for best sunlight hits.

Hint: Leave the AC plug plugged in at all times. It will not draw power until you throw switch S1 onto "charge."

Table 17. Parts List for Solar TV Reception Booster.

Item No.	Description
C1	56-pF capacitor.
C2, C3, C6	.0022-μF capacitors.
C4	1.5-pF capacitor.
C5	18-pF capacitor.
C7	150-μF, 15-w VDC electrolytic capacitor.
C8, C9	.0022-μF capacitors.
L1	7½ turns No. 24 enameled wire, evenly wound on a 3/16-inch-diameter form.
L2	3 turns No. 24 enameled wire evenly wound on a 3/16-inch-diameter form.
L3	17 turns No. 24 enameled wire evenly wound on a 3/16-inch-diameter form.
L4	11½ turns of No. 24 enameled wire evenly wound on Speer Type E ferrite form.
L5, L6	10 turns No. 24 enameled wire evenly wound on Speer Type E ferrite form.
M1	4½ VDC, with three penlites of NiCd cells.
M2, M3	Binding posts.
M4	AC wall plug with cable clamp.
M5	4½-VDC Solar pack. (International Rectifier No. SP5G26C or equiv.)
Q1	HEP-3 transistor.
R1	51K resistor.
R2	30K resistor.
R3	240-ohm resistor.
R4	110-ohm resistor.
R5	750-ohm resistor.
S1	Spst switch.
T1, T2, T3, T4	Using Ferroxcube K5050-06 ferrite cores, insert 2 turns of special 300-ohm miniature twinlead in each form hole. Pull tight, and connect leads where shown in diagram.
T5	6.3-VAC filament trnasformer. (Triad F-14X or equiv.)

18

Free-Power AM Radio Receiver

Strange as it may seem, this transistor broadcast-band radio receiver "steals" power from one station to give to another! The principle is basic: By tuning the battery-section antenna coil (L2) to the strongest broadcast station on the band, diode X1 can rectify the rf and convert it into DC current. Naturally, the closer you are to a strong station, the more current the "radio battery" section of your radio receiver will be able to supply. Once you have found this spot, the DC current is passed on to power the transistor circuit which acts as a genuine receiver, with the full tuning it affords. See Fig. 18 and Table 18.

The basic consideration is a good antenna and ground, the latter preferably being made to a water pipe or solid external ground composed of a pipe driven at least 4 feet into moist earth. This procedure not only ensures maximum signal pickup for the radio-battery portion of the circuit, but also provides best results for the GE-2 receiver circuit.

Once completed, just tune the radio battery as explained in the first paragraph and calibrate your receiver by adjusting L1 so that the bottom of the band occurs when C1 is fully meshed. If you have a local broadcast station operating near 540 kHz, this simplifies things tremendously. Once the calibration procedure is complete, forget entirely about adjusting L1 and do all your listening by tuning C1.

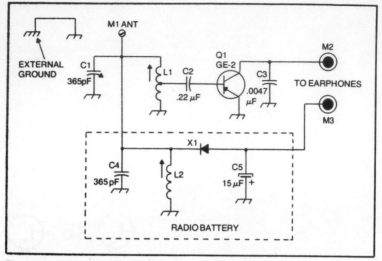

Fig. 18. Free-power radio receiver.

Table 18. Parts List for Free-Power AM Radio Receiver.

Item No.	Description
C1, C4	365-pF variable capacitors.
C2	.22-μF capacitor.
C3	.0047-μF capacitor.
C5	15-μF, 6-w VDC electrolytic capacitor.
L1, L2	Loopsticks. (Superex VLT-240 or equiv.)
M1	Binding post.
M2, M3	Test jacks.
Q1	GE-2 transistor.
X1	1N38B diode.

19

Electronic Stethoscope

Have you seen the ads by surveillance equipment companies offering "ultra-sensitive electronic stethoscopes" for prices ranging from $100.00 to upwards of $200.00? The circuit shown in the accompanying diagram will do everything those jobs will, and for less money.

Probably the biggest-selling wall-listening devices on the underground eavesdropper market, these audio stethoscopes enable the user to pick up voices clearly through thick walls and record them by simply feeding the output (shown as going to a speaker in the diagram) capacitively to a tape machine. Originally designed by a Philadelphia doctor who was losing his hearing, these gadgets are now being used by the medical profession for a number of purposes.

You can build yours in whatever fashion you like, just taking care to follow the schematic closely, particularly with regards to proper hook-up of connections to the audio amplifier (M1). The microphone you choose can be a standard contact mike such as manufactured by Telephone Dynamics (such as the Vibra-Mike) or any one of a number of conventional brands sold through Allied Radio, or Radio Shack, etc.

If you would rather be less fancy, you can reverse-wire a standard transistor radio type earpiece for operation as the eavesdropper microphone, simply taping it to the wall securely for maximum pickup. Volume to the speaker is controlled by potentiometer R1. See Fig. 19 and Table 19.

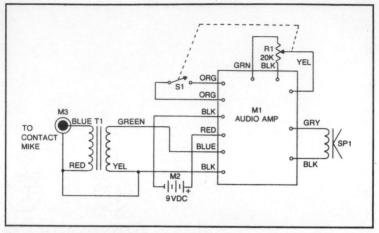

Fig. 19. Electronic stethoscope.

Table 19. Parts List for Electronic Stethoscope.

Item No.	Description
M1	Amplifier module. (Lafayette 99 C 9037 or equiv.)
M2	9-VDC battery.
M3	Microphone receptacle.
R1	20K potentiometer with spst switch.
S1	Spst switch on R1.
SP1	8-ohm speaker. (Lafayette 99 T 6036 or equiv.)
T1	Audio transformer. (Lafayette 99C 6034 transformer or equiv.)

20

Bathtub Overflow Alarm

How many times have you left the bathtub faucet running when you left to answer the telephone, only to return to 3 or 4 inches of water on the bathroom floor and a leaky ceiling downstairs? If this has happened to you as frequently as it has to us, you would do well to take a close look at the simple circuit in Fig. 20.

When the water reaches a predetermined level, a howling shriek will spew forth from the speaker, signaling for some distance and you had better get back fast and turn the water off. Since the water completes the circuit of the alarm, there is no danger of shock, although it's a good idea to remove the sensor once it has done its job. If you like, you can add an spst switch between M1 and R1 which, when opened, will shut the alarm circuit off. Once this is accomplished, not the most remote chance of shock remains, provided you have taken pains to ensure that the alarm itself is well-housed in an insulated (preferably plastic) box.

The sensor should be entirely homebrew and consist of two stiff wires, such as clothes hanger wire, that have had the insulating paint scraped off and can be mounted on a rather circular lip which will fit securely over the rim of the bathtub. The lip can be made out of any good nonconducting material so long as it has the flexibility required for "forming" it into shape. The authors used a stiff piece of plastic box, placed it over the bathtub rim, and heated it until the plastic went limp. Once hardened, it was trimmed and formed the mounting plate, or lip.

As suggested in the illustration, a good housing for your alarm circuitry is a plastic speaker box similar to those being offered by the major parts wholesalers. Also, see Table 20.

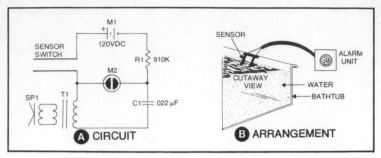

Fig. 20. Bathtub overflow alarm.

Table 20. Parts List for Bathtub Overflow Alarm.

Item No.	Description
C1	.022-μF capacitor.
M1	120-VDC battery.
M2	NE-2 bulb.
SP1	Transistor radio speaker, 3.2 ohms.
T1	Transistor radio speaker transformer.
	One clothes hanger.
	Plastic speaker box.
	Plastic box.

21

Aquarium Heater Silencer

Plagued by reports of neighborhood TVI and BCI, one of the authors noticed that the interference complaints persisted even when he was not on the air. In one case, on return from a two-week vacation the next-door neighbor insisted that the author had left a transmitter running, because he continued to get interference, particularly annoying to him on the AM broadcast band. This problem sounded like a power-line problem, so the local power company was called onto the scene.

Imagine the surprise to learn that the trouble emanated from an aquarium heater in a tropical fish tank a block away! Apparently the heater arced over just before it came on, and just after it went off, in its normal thermostatic operation. This arcing, traveling through the house wiring, found its way into the main street lines, and worked its way into every home on the block.

Unfortunately the party with the cheap aquarium heater did not want to invest in a nonarcing model. Hence, the circuit shown in the accompanying diagram was devised. Even with brand-new parts, it should not cost over $5.00, including the minibox housing.

Functionally, the silencer filters out "popping" noises by grounding them *before* they reach the main power line. In operation, you merely plug the silencer into the nearest AC outlet, and plug the aquarium heater cord into the silencer socket.

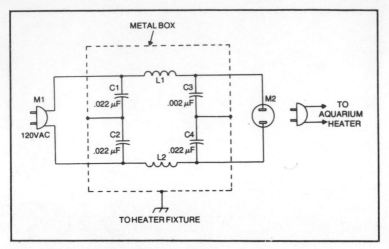

Fig. 21. Aquarium heater silencer.

Table 21. Parts List for Aquarium Heater Silencer.

Item No.	Description
C1, C2, C3, C4	.022-μF capacitors.
L1, L2	Solid windings of No. 16 enameled wire on 1-inch diameter form. Fill with solid windings to a 5-inch overall length.
M1	AC wall plug, with cap and cable clamp.
M2	AC socket.

56

22

Electronic Tic-Tac-Toe

Here is a dandy project that requires only a good supply of 6.3 volt bulbs, a filament transformer, assorted spst switches, an spst switch, plywood board, and a supply of hook-up wire to build an impressive automatic electric tic-tac-toe machine which will fascinate the youngsters and oldsters alike for many hours on rainy days.

The secret of this game, which is built on the back of a sheet of plywood which is painted with cross-hatch tic-tac-toe lines, is the use of colored filament bulbs. The reason for the nine green bulbs and nine red bulbs is so you can keep track of which move was made by which player. Simply assign one player one color, and the other player the other color. Incidentally, you do not *have* to go searching for colored light bulbs; you can just as well drill "peep holes" through the tic-tac-toe board and tape Kodachrome transparency plates across the holes on the game side so that the light streaming through will take on the hue of the transparent sheet. In any case, you will have to arrive at a color arrangement in order to keep from getting one player's moves confused with another's.

With the switches wired as shown in Fig. 22, each of the nine switches corresponds to one of the tic-tac-toe boxes on the crosshatch pattern. Whether you throw the switch right or left to go on is determined by which light you want to become illuminated. For example, the "red" player always switches right. The "green" player switches left. It is impossible to make both bulbs in any given box come on at one time.

When the game is complete and winner established, merely return all switches to a center position and all the bulbs will go out. Now you are ready for your second game. Also, see Table 22.

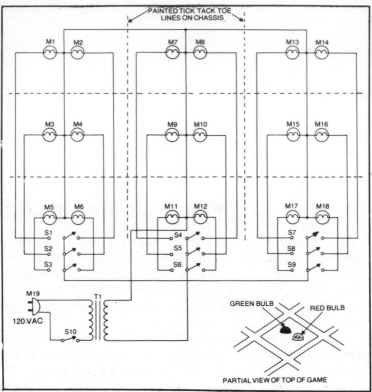

Fig. 22. Electronic tic-tac-toe.

Table 22. Parts List for Electronic Tic-Tac-Toe.

Item No.	Description
M1, M3, M5, M7, M9, M11, M13, M15, M17	6.3-VAC green pilot bulbs.
M2, M4, M6, M8, M10, M12, M14, M16, M18	6.3-VAC red pilot bulbs.
M19	AC wall plug, cap, and cable clamp.
S1, S2, S3, S4 S5, S6, S7, S8, S9	Spdt toggle switches (center: off).
S10	Spst switch.
T1	6.3-VAC filament transformer. (Triad F-14X or equiv.)
	Plywood slab about 1 ft by 1 ft.

23

Fence Charger

Believe it or not, this circuit generates *thousands* of volts of electricity. All you need is a GE-3 transistor, a standard 6.3-volt filament transformer, a handful of spare parts, and a metal fence.

As you might suppose, the transformer is the key to the operation, in addition to providing isolation from the "charged" circuit. This way, if something should short out a quarter mile away, you won't burn up your charger. The amount of shock you wish to send down the fence is determined by the setting of potentiometer R1, a 15,000 ohm variable resistor. For maximum shock, set R1 at maximum.

Incidentally, although this circuit is portable in nature and is powered entirely from a hefty 6-volt source, care should be taken if the circuit is "on" and small children are in the vicinity. The current rating is quite low, but a child's resistance to sustained high-voltage sources (such as if he were caught in the fence with his feet on the ground) is nowhere near that of an adult.

Your earth ground at point of feed should terminate in a 4-foot ground pipe near the fence itself. The fence should be inspected to ensure that at no point does the wire come in contact with shrubbery or ground. Next, the ground near the entire length of electrified fencing should be moistened from time to time so the ground connection carries the full length. If the ground is extremely dry, the shock effect will dissipate with distance away from the feed point.

59

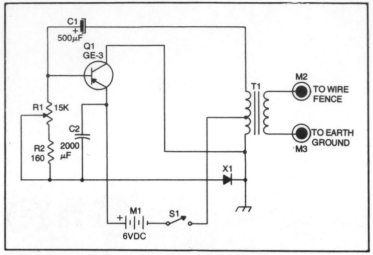

Fig. 23. Fence charger.

Check the battery condition from time to time, as well as outdoor connections. Rust and dirt can ruin the best electrical joints if you don't inspect them periodically. See Fig. 23 and Table 23.

Suggestion: If your local laws do not prohibit it, this circuit is great for a portable shocker for warding off attachers. Use flashlight cells and build it with a portable "shock rod" with two prongs, one for the "hot" lead and one for "ground." NOTE: In New York a young woman lost a legal case because she had employed a shock rod as her defense when attacked by a night assailant. The shock rod, while not illegal to possess, is *illegal* to use in Manhattan.

Table 23. Parts List for Fence Charger.

Item No.	Description
C1	500-μF, 10-W VDC electrolytic capacitor.
C2	2000-μF, 15-w VDC electrolytic capacitor.
M1	6-VDC battery.
M2, M3	Test jacks.
Q1	GE-3 transistor.
R1	15K potentiometer.
R2	160-ohm resistor.
S1	Spst switch.
T1	6.3-VAC filament transformer (Triad F-14X or equiv.)
X1	1N540 diode.

60

Receiver Signal Alarm

Have you ever waited hours on end for a certain call to come across your CB set, afraid to leave the room for fear you would miss it? Well, this gadget will set off a bell or buzzer alarm whenever it "hears" an encoded signal on the receiver. This is especially good for ham radio operators belonging to a 24-hour-per-day network on 2 meters, and for monitoring emergency CB channels. As the call comes through the high-pitched "beep" tone, the buzzer "fires."

Interestingly, a door-chime is particularly effective. The authors set one up next to the original front-door bell ringer and wired it to the CB set (which was fixed-tuned to CB Channel 2, a local emergency frequency). It has a different tone altogether from the regular doorbell, yet it is pleasing to hear and not at all bothersome. It takes its power from the original bell transformer in the basement (24 VAC) and is switched on and off through the receiver signal alarm box.

NOTE: You do not have to disconnect your speaker; just wire in the circuit in Fig. 24 across the two speaker cables and adjust R3 for the desired audio level required to "kick-in" the alarm circuit. Also, see Table 24.

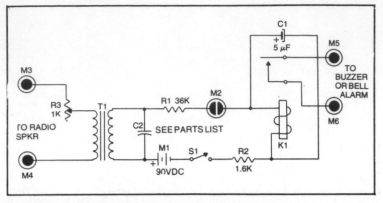

Fig. 24. Receiver signal alarm.

Table 24. Parts List for Receiver Signal Alarm.

Item No.	Description
C1	5-μF, 100-w VDC electrolytic capacitor.
C2	.001-to 1.0-μF capacitor.*
K1	2.5K relay. (Sigma 4F2500-S/SIL or equivalent.)
M1	90-VDC battery.
M2	NE-2 bulb.
M3, M4, M5, M6	Test jacks.
R1	36K resistor.
R2	1.6K, 1-watt resistor.
R3	1K potentiometer.
S1	Spst switch.
T1	Audio output transformer. (Stancor A-3327 or equiv.)

*Experiment with value for best performance. (Value relates to frequency of beep on signal being received, e.g., for higher-pitched tones, lower values work better.)

⚛ 25 ⚛

Magic Magnet Box

Here is a project that has been amusing the children of one of the authors for months now and has never ceased to amaze their friends. It is known in his family as the magic magnet box and only the older boy knows how it works. To the rest, it's a complete mystery. Place any metallic object on top of the box, hit switch S1, and watch the action. Metal toy cars scurry about, miniature metal soldiers march in drill-like fashion, and odd metal objects (like paper clips and nuts and bolts) spin incessantly at a predetermined rate.

The secret, of course, is in the cylindrical magnet mounted on top an inexpensive, 1½-volt DC motor. The magnet is situated just below the top of the wooden box and sets up wild force fields when it is set into motion by the small motor.

If you like, you can put a knob on R1 and combine S1 for single-knob control. Switching on the control starts the motor; advancing the R1 setting determines how fast the magnet will spin. See Fig. 25 and Table 25.

Incidentally, the use of painted metal filings on a stark red or yellow paper laid on top the box provides interesting psychedelic patterns.

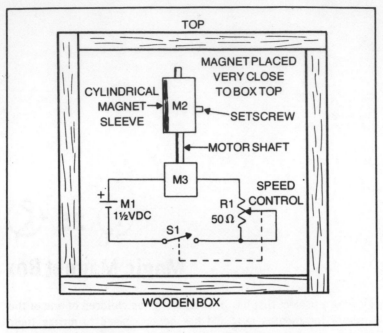

Fig. 25. Magic magnet box.

Table 25. Parts List for Magic Magnet Box.

Item No.	Description
M1	1½-VDC cell.
M2	Cylindrical magnet. (Edmund Scientific No. P-40,418 or equiv.)
M3	Battery-operated electric motor. (Edmund Scientific Co. No. 30,305 or equiv.)
R1	50-ohm potentiometer.
S1	Spst switch.
	Wooden box.

26

Electric Combination Lock

Here is a circuit, limited only by your imagination, that will guarantee secrecy and privacy, as well as baffle the most determined would-be intruders. It is a complex electric locking system which depends entirely on a straight-through electric circuit for it to function.

It can be used "as is" for a number of applications. For example, you can simply feed the input to the household AC line and the output to the 120-VAC line going to your ham transmitter, CB rig, TV set, or the like. If the lock box is permanently installed as part of the wiring between the AC outlet and appliance, radio, or whatever, it will be impossible to get the unit to function without knowing the secret combination.

In the circuit shown in Fig. 26, the switches are placed *exactly as shown* on the face of a panel. This means that there will be two control knobs (each of the switches used will "read out" to numbers 1-13 on your panel) in the top row, two in the next row, and the last positioned on the bottom between the two above. It looks as if you want to start at the upper-left; actually it would take you about three weeks of experimenting, starting from the bottom switch, to unlock the mystery. In the illustrated circuit, the combination, reading from S1 to S5, is 4-11-1-9-13. Unless all switches are dialed in order to those numbers, nothing will happen.

Additionally there is the hidden switch (S6) which must be depressed *after* all the preceding has been performed in order for the circuit to operate. It's best to hide this switch in a drilled-out recess and paint it over so as to conceal it entirely. Only you will know where it is. If you do recess it, it will take the eraser end of a pencil to depress it, particularly if you use a tiny, subminiature type.

If you have a laboratory, electronic workbench area, or other room which you do not want open to intruders, buy a *solenoid-operated door latch* through your local hardware store and replace the conventional door lock with this mechanism. Some operate on as little as 6 VDC, others on standard household current. The electric combination lock can be installed on the door itself, with the knobs and numbers projecting on the *outside* (facing away from your room) of the door.

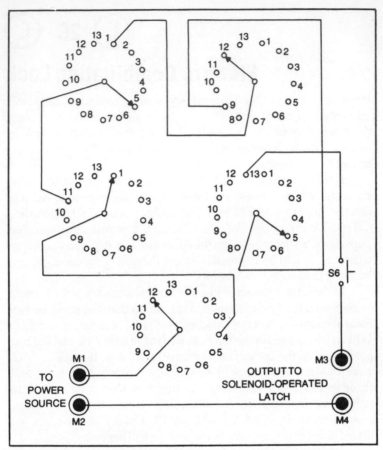

Fig. 26. Electric combination lock.

Incidentally, in the five-switch circuit there are 371,293 possible combinations. If you use a lower number of switches, here are the numbers of combinations: 2 switches, 169 combinations; 3 switches, 2197 combinations; and 4 switches, 28,561 combinations.

Table 26. Parts List for Electric Combination Lock.

Item No.	Description
M1, M2, M3, M4	Test Jacks.
S1, S2, S3, S4, S5	13-position TV channel switches (rotary).
S6	Push-button switch, normally open.

27

Echo Box

Ever wish you had a professional echo chamber similar to those used by disk jockeys? All you need is a tape recorder, an expensive replacement tape head, a 2N417 transistor, and handful of components. Construct the circuit carefully, and you will sound like you are in Grand Canyon. Figure 27 (C) shows how the echo box is hooked into your audio or transmitting system; the pictorial (B) shows how the echo effect is obtained through installation of a new recording head. Also, see Table 27.

If you want to be able to vary the spacing of the echoes, you will want to be able to "slide" the new tape head back and forth to obtain this feature. Naturally, the farther away you place the new tape head from the original head, the greater this spacing will become. Adjust the "dominance" through potentiometer R5, which permits your allowing the echo to be the primary audio source, or the original voice to be the primary audio, whichever you prefer. NOTE: Although the schematic shows the output going "to volume control on amp," the output can be coupled to the input of most any amplifier. If you have a stereo amplifier, you can run the main recorder output to channel 1, and the echo box output to channel 2. The effect is interesting.

If you have no additional amplification requirement, then wire into the "hot" lead of the volume control on the recorder itself. This permits your playing back any recording you make, inserting the echo effect by simply adjusting the dominance control.

If you want to add the echo effect to your CB or ham transmitter, feed both recorder and echo box outputs together, so that you wind up with just one "hot" and one "ground" lead. Preferably this should be accomplished through the use of common garden-variety microphone cabling. Just before you plug this single mike cable into your mike connector on the CB rig, insert a 15-μF capacitor in the "hot" lead. This capacitor serves to isolate the output of the recorder/echo box from the modulator to avoid

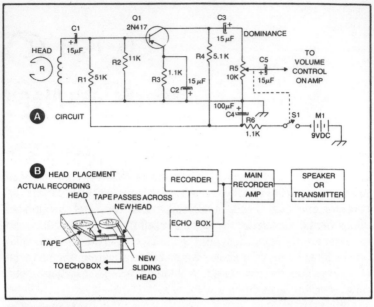

Fig. 27. Echo box.

damaging the transmitter. If you do feed it into the modulator, leave the modulation gain control exactly where it was originally set for straight mike operation, and do your level-adjusting with the volume control on the recorder and the dominance potentiometer (R5). You will not need much output, since the audio amplifier in the modulator does most of the work. Keep levels of both controls down, monitoring output through a crystal detector or grid-dip meter with earphones to avoid distortion.

Table 27. Parts List for Echo Box.

Item No.	Description
C1, C2, C3, C5	15-μF, 50-wVDC electrolytic capacitors.
C4	100-μF, 50-wVDC electrolytic capacitor.
M1	9-VDC battery.
Q1	2N417 transistor.
R1	51K resistor.
R2	11K resistor.
R3, R6	1.1K resistors.
R4	5.1K resistor.
R5	10K potentiometer with spst switch.
S1	Spst switch on R5.
	Tape recording head. (Lafayette Radio 99R 6194 or equiv.)

28

Pool Splash Alarm

If you are the proud owner of a backyard swimming pool, chances are that you are also the owner of one long continuous headache. The problem with these pools is keeping the neighbor children away when you are not around, plus animals and small children who might wander near the pool at night, creating the danger of drowning in an unsupervised swimming pool. While the pool splash alarm described here is no panacea, it does help, particularly at night when the household is asleep. Its drawback is that it responds only to a sudden wetness resulting from a splash, and this is something to think about before you go advertising your new invention all about the neighborhood. You will be somewhat aggravated when you find out how many stones get thrown in your pool at night by curious youngsters eager to see your lights come on and the buzzer blare.

So do not tell everyone what you are doing. You can simply describe the probe wires as a gadget that has something to do with the water level and purifying mechanisms and leave it at that.

The main splash alarm circuitry should be in a box somewhere in the house or garage, with insulated cabling to the probes either buried underground or held down with clamps to prevent someone from tripping on them.

The probes can be separated by any length at all. One must be under water at all times, while the other has to be adjusted ever so carefully to detect a sudden rise in water level such as would occur by someone falling (or slipping) into the pool. In permanent installations, it is advisable to use a strip of conductive paint on the side of the pool just ½ inch above the water line, although this requires that you maintain a predetermined water level in the pool at all times. For other installations a stiff wire probe (such as No. 10 or No. 12 bare wire) can extend down over the rim of the pool to a point just above the water surface. If you design it correctly, you will be able to adjust the height of the probe from time to time. See Fig. 28 and Table 28.

The authors suggest that your buzzer be placed in a position where it can be heard in all occupied sections of the house. If this is

69

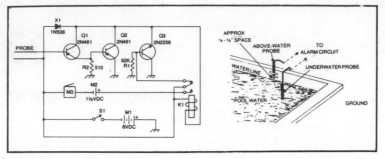

Fig. 28. Pool splash alarm.

impossible, string several buzzers in series including one in the hall just outside the master bedroom. Test the probe setting frequently to ensure that wind hitting the water will *not* activate the circuit but that a small youngster getting into the pool *will*. Open switch S1 whenever the pool is in use.

For those new to backyard pools who might be tempted to wire this to an outdoor clanging bell, forget it. If you are not at home (or away on vacation), this will not necessarily bring neighbors to the aid of the fallen child or animal. Your responsibility is to drain the pool whenever you are not going to be at home, or to have it securely covered. Neither insurance firms nor the law recognizes pool splash alarms, and ultimate responsibility reverts to you.

Table 28. Parts List for Pool Splash Alarm.

Item No.	Description
K1	6-VDC spdt relay. (Potter & Brumfield RS5D or equiv.)
M1	6-VDC battery.
M2	1½-VDC cell.
M3	1¼-VDC buzzer or alarm bell.
Q1, Q2	2N461 transistors.
Q3	2N2256 transistor.
R1	62K resistor.
R2	510-ohm resistor.
S1	Spst switch.
X1	1N536 diode.

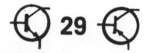

Wireless Mike

Here is a dandy little home broadcaster that you can carry about or use as a fixed-station transmitter, yet it can be built for little or no cost. It operates in the broadcast band and will carry for a considerable distance to any AM radio.

It can be constructed to fit into a tiny plastic box with holes drilled in the side to permit walkie-talkie type operation to a self-contained microphone, or equipped with a mike jack to accept music programming from a record player (in which case you should install a 15-μF capacitor in the "hot" output lead of the player if you're not taking your output directly from the cartridge) or from any conventional ceramic or crystal microphone. Calibrate the transmitter by the same method used in earlier AM projects (adjust L1 for maximum spread across the whole of the band by C1). See Fig. 29 and Table 29.

In operation, tune the AM receiver to a spot not occupied by any commercial station and zero in by tuning C1. The transmitting range is determined by the overall length of the antenna. NOTE: FCC Rules & Regulations, Part 15, allow only for a maximum transmit distance of 300 feet. Exceed this by inductively hooking into phone lines or long-wire antennas, and you will not only be heard all over town, but you will have the FCC after you.

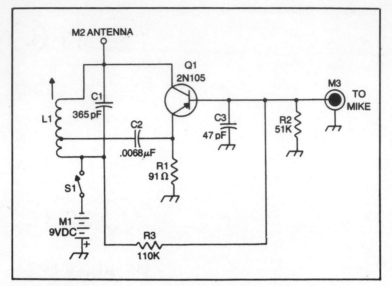

Fig. 29. Wireless mike.

Table 29. Parts List for Wireless Mike.

Item No.	Description
C1	365-pF variable capacitor.
C2	.0068-μF capacitor.
C3	47-pF capacitor.
L1	Tapped loopstick. (Lafayette 32 H 4108 or equiv.)
M1	9-VDC battery.
M2	Binding post.
M3	Microphone receptacle.
Q1	2N105 transistor.
R1	91-ohm resistor.
R2	51K resistor.
R3	110K resistor.
S1	Spst switch.

30

Automatic Animal Flasher

Are you a photographer who likes to get good out-of-doors pictures of wild animals in their natural habitat? Ordinarily, such photos are beyond the abilities of the average camera enthusiast, but with the flasher device described here, you will be turning out amazing pictures on a par with the professionals.

The key to good night shots is catching the animal totally unaware. Aside from sitting in the woods all night freezing as you wait for raccoons and deer who know better, you can simply rig your camera up on a pole, wire in your flasher system, and plant the "bait" (which can be anything you think will attract wildlife, such as bread crumbs, jelly, etc). The automatic flasher will "sense" any animal activity and immediately get off a shot before the creature knows what hit him.

As shown in Fig. 30, this circuit has been set only to activate a flashbulb mounted in such a way as to throw maximum illumination upon the animal sniffing around the food supply. This means you'll either have to wire in the shutter mechanism to the circuit, or follow the conventional method of opening the shutter to a setting or exposure correct for the flash being furnished. The lens will have to stay open until you retrieve your camera or reset the system. This is not really a problem, however, since the flashbulb will fire and then go out, leaving the camera in near-total darkness. A double-exposure is impossible unless the camera is left out all night until the sun comes up.

As indicated in Fig. 30, three sensing elements are used, although you can elect to use any one, or two, or combination to achieve maximum animal sensitivity. The microphone is by far the most dependable of them all, although it might not respond to *every* presence, particularly of certain birds. In that case, the photocell can be employed by directing a stream of light at the cell. If a bird or other animal interrupts that light stream by casting a shadow on the photocell, the flashbulb will fire. The moisture-sensitive sensor plate is a conventional type available through most large electronics and scientific supply companies. Be sure, however, you buy the most sensitive plate available, *not* the kind designed to sense only large drops of rain. An animal's feet are normally much

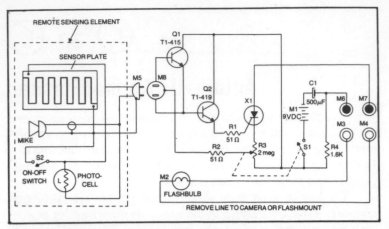

Fig. 30. Automatic animal flasher.

more damp than human skin; hence when the animal steps on the plate enough of an electrical connection will be made to trigger the flasher circuit. WARNING: Do not use the sensor in extremely damp weather.

The sensitivity you want is determined by where you set R3. Maximum sensitivity may be too great; it might respond, for example, to the sound of birds fluttering in overhead trees. Try to imitate the intensity of the sound you think the animal will make, and use a homebrew cardboard tube or parabola-type cone over the mike to direct its pickup directly at the baited spot. You'll lose a few flashbulbs in setting it up, but your photos will be prizewinners.

Table 30. Parts List for Automatic Animal Flasher.

Item No.	Description
C1	500-μF, 25-wVDC electrolytic capacitor.
M1	9-VDC battery.
M2	Standard flashbulb.
M3, M4	Single-prong test plugs.
M5	2-conductor shielded microphone plug.
M6, M7	Test jacks.
M8	2-conductor shielded microphone receptacle.
Q1	TI-415 transistor.
Q2	TI-419 transistor.
R1, R2	51-ohm resistors.
R3	2-megohm potentiometer with spst switch.
R4	1.6K resistor.
S1	Spst switch on R3.
S2	Spst switch.
X1	2N3228 SCR.

31

Long-Distance
Light-Ciphering Machine

Recall the flashlight radio transmitter described in Project 16? This light-ciphering system is much less complicated, is somewhat more dependable over a greater distance, and is less expensive if you are buying all new parts. Communications are carried out by tone beeps over a Sonalert generator, a great gadget for devices which operate on small voltages. You can use International Morse Code (available in ARRL publications, as well as Boy Scout manuals and other sources), or you can simply make up your own beeping-code system, using certain arrangements of long beeps and short beeps to represent secret prearranged codewords.

Unlike the other project, no magnifying lens is required in this system. Additionally, you can either use the push button on the primary flashlight on-off switch as your signalling device (taking care that you do not move your light beam off target), or you can wire it for outboard operation as shown in Fig. 31. In any case, the signal light is not part of the base code-ciphering circuit and consequently separate battery supplies are provided. For best results, you will probably want to separate the two major items—the photocell "target" and the flashlight signaller—somewhat in the final physical layout of your system.

The only thing you have to worry about is shielding the photocell from unwanted sources of stray light. If you build this as a permanent communications system between, for example, your house and your friend's down the street, you can mount both the flashlight and photocell remotely on the roof or on a ledge placing them in direct line-of-sight with each other and preaiming both systems. Once the sensor and flashlight have been secured in

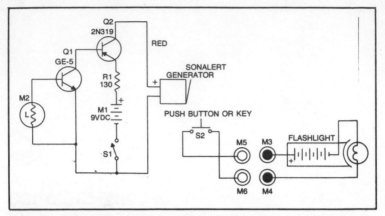

Fig. 31. Long-distance light-ciphering machine.

position, cables can be run down to your base ciphering center anywhere in the house, where you can send and receive in relative comfort from any desk or table-top where you have elected to set up your equipment. Should you do this, you'll probably want to use an inexpensive Morse Code key to activate the rooftop flashlight.

Incidentally, the brighter the light you use, the more distance you'll have. Once the sun goes down, you're ready to operate. By the way, this is one of the few inexpensive light-ciphering systems going that permits *simultaneous* send and receive.

Table 31. Parts List for Long-Distance Light-Ciphering Machine.

Item No.	Description
M1	9-VDC battery.
M2	Photocell. (International Rectifier S1M or equiv.)
M3, M4	Single-prong test plugs.
M5, M6	Test jacks.
Q1	GE-5 transistor.
Q2	2N319 transistor.
R1	130-ohm resistor.
S1	Spst switch.
S2	Push button or key.
	Flashlight.
	Sonalert SC-628 sound generator or miniature buzzer with minimum of current drain.

32

Adjustable Electronic Flasher

We have all seen the four-way flasher systems installed on automobiles. Although not required on older and second-hand vehicles, these flashers are extremely useful, particularly when your car is stalled on a highway at night because of a flat tire or some other reason. If you buy a packaged four-way flasher add-on unit from an auto supply store, you will note that it is designed so as to flash your taillights. The circuit shown in Fig. 32, however, can be adapted to flash any 12-volt lamps you want—including your main sealed-beam headlamps up front.

The unit can be built in a small plastic box or metal enclosure and simply bolted in place against the firewall in the engine compartment. Switch S1 should be a remote "on-off" toggle switch mounted just beneath the dashboard. Take care in wiring the lamps that you only use those bulbs on a circuit together in series as demonstrated by the placement of M2, M3, M4, and M5 in the schematic.

NOTE: This unit has been designed only for 12-volt negative ground cars. Also, see Table 32.

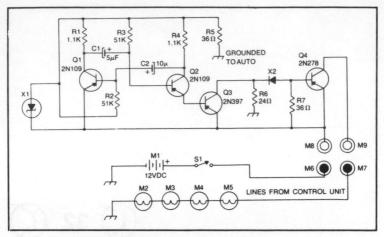

Fig. 32. Adjustable electronic flasher.

Table 32. Parts List for Adjustable Electronic Flasher.

Item No.	Description
C1	5-μF, 12-w VDC electrolytic capacitor.
C2	10-μF, 12-w VDC electrolytic capacitor
M1	Auto (car) battery.
M2, M3, M4, M5	Auto taillights or similar electric bulbs.
M6, M7	Single-prong test plugs.
M8, M9	Test jacks.
Q1, Q2	2N109 transistors.
Q3	2N397 transistor.
Q4	2N78 transistor.
R1, R4	1.1K resistors.
R2, R3	51K resistors.
R5, R7	36-ohm resistors.
R6	24-ohm resistor.
S1	Spst switch.
X1	10-volt, 1-watt zener diode.
X2	500-mA, 200-PIV rectifier.

33

Matchbox Oscillator

Here is an audio oscillator, CPO, or just plain beep-box which you can build—almost literally—into a matchbox. It requires only eleven components plus an 8-ohm subminiature transistor radio type speaker to yield a penetrating blast of tone to suit a multitude of purposes.

The heart of the tiny oscillator is the GE-5 and 2N105 configuration which requires only a subminiature hearing-aid type battery supply to set the circuit into full operation. There is no need for a matching transformer; the speaker is connected directly to the output.

Switch S1 can be a push button as shown in Fig. 33 or take the form of a more permanent control, such as a standard spst toggle affair. If a cheap code practice oscillator is what you have in mind, substitute a code key for S1. No current is drawn by the oscillator except when the switch is thrown; hence, there is no need for other controls except to vary the tone. This is accomplished through variable resistor R2, a 75K potentiometer.

Although other values can be substituted for the .047-μF capacitor (C1), the authors found that this particular capacitor provides the greatest range of tone possibilities in conjunction with R2, the "tone adjust" control.

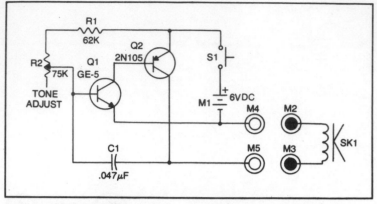

Fig. 33. Matchbox oscillator.

Table 33. Parts List for Matchbox Oscillator.

Item No.	Description
C1	.047-μF capacitor.
M1	6-VDC hearing-aid subminiature battery supply.
M2, M3	Single-prong test plugs.
M4, M5	Test jacks.
Q1	GE-5 transistor.
Q2	2N105 transistor.
R1	62K resistor.
R2	75K potentiometer or variable resistor.
S1	Push-button switch.
SP1	Transistor radio 8-ohm speaker. (Lafayette 99 T 6032 or equiv.)

FM Music Receiver

Here is a clever one-transistor FM receiver which is well worth your time and effort. It will do an adequate job in most metropolitan areas, and it exhibits an extraordinary ability to pull in the weak stations if you live out in the country. If you start from scratch, with all-new components, this project will not cost more than a few dollars.

The heart of the receiver is an efficient superregenerative design and a "gimmick" capacitor linked between the emitter and collector of Q1. This capacitor is really a homebrew type you can make from two 1-1/16-inch lengths of No. 16 insulated hookup wire. Just twist them together for best results. Anywhere from ½ to 4 turns should do the trick.

Adjust the potentiometer until a "rushing" sound is heard in the earphones—right past the threshold point—and you are all set. Tuning is accomplished with C2. See Fig. 34 and Table 34.

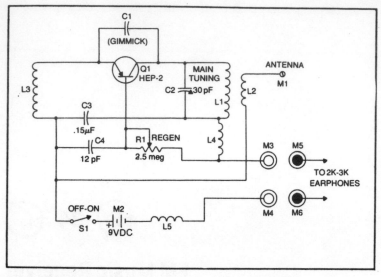

Fig. 34. FM music receiver.

Table 34. Parts List for FM Music Receiver.

Item No.	Description
C1	"Gimmick" capacitor: ½ to 4 turns No. 16 insulated hookup wire. (See text.)
C2	2-30-pF variable capacitor.
C3	.15-μF capacitor.
C4	12-pF capacitor.
L1	8¾ turns of No. 20 wire, ¼-inch diameter, ¾ inch long.
L2	2¾-turn link of No. 20 wire, spaced ⅛ inch from L1.
L3, L4, L5	7-μH coils. (Ohmite Z-50 or equiv.)
M1	Binding post.
M2	9-VDC battery.
M3, M4	Test jacks.
M5, M6	Single-prong test plugs.
Q1	HEP-2 transistor.
R1	2.6-megohm potentiometer.
S1	Spst switch.

35

Transistor Tickler

Are you one of those types who just cannot pass up an opportunity for a little fun? The mysterious tickler box is just the project for you. You merely hand it to someone and stand back to watch the action. Our bet is he will drop it like a hot potato, unless he is unusually accustomed to receive jolts of electricity in the *thousands*-of-volts range!

Actually, this circuit is nearly harmless, since the 1½-volt flashlight battery can hardly supply enough current to give more than the feeling of a slight, but annoying, tickle to whoever comes in contact with the tickler/sensor. Yet it's great for parties, particularly when you tell your victim that he's just received a mild 20,000-volt shock!

The heart of the circuit is the filament transformer, which does most of the dirty work. For best effect, the tickler should be constructed in a cardboard or plastic box with the sensing/tickling element wrapped around the outside of the box. One way to do this is simply to wind bare coil wire around the box in such a way as to insure that the coils do not come in touch with one another, yet will be hard to hold without "shorting" them out with the skin of the victim's hand.

A much simpler way to build this gadget is merely to fasten two metal plates to the box, one to cover the bottom of the box, and one for the top. The top plate can also extend over the sides, so long as it does not electrically touch the bottom plate. These plates can be made out of any flexible metal sheeting (even aluminum foil) and fastened to the box by machine screws or glue. The two wires from the output of the filament transformer are connected, on each, to the plates. In this manner it is virtually impossible for someone *not* to receive a tickle when he picks up the box.

Incidentally, this upper-and-bottom-plate concept works well in actual use. If you build it for a youngster, for example, you can

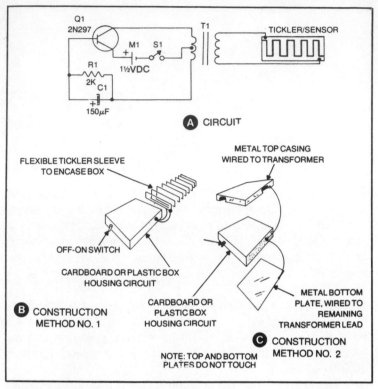

Fig. 35. Transistor tickler.

show him how to hold the device without coming in contact with both metal plates at one time. It takes some time before his friends catch on to why he seems impervious to the transistor human tickler. See Fig. 35 and Table 35.

Table 35. Parts List for Transistor Tickler.

Item No.	Description
C1	150-μF, 15-w VDC electrolytic capacitor.
M1	1½-VDC cell.
Q1	2N297 transistor.
R1	2K resistor.
S1	Spst switch.
T1	6.3-VAC filament transformer. (Triad F-14X or equiv.)
	Cigar or similar-type box.
	Wire or metal plates.

36

The "Chaos-Pandemonium" Box

This project is possibly the worst thing you could do to someone. Once it has been received by your "victim," nothing he does will shut it off. It screams, it wails, it flashes lights perpetually, all the while appearing to be broken. If he pushes the "panic" button, it only makes the siren start. If he throws the on-off control into the "off" position, the siren will stop, but lights will start flashing. If he tries to turn off the lights by returning S2 to the "on" position, the siren wails once more, although the lights go out. Unless he knows the secret location of the concealed switch S3, he'll never stop the infernal machine.

Incidentally, depending on the temperament of your victim, you can preset the audio output up to the point of ear-splitting volume. The three-transistor amplifier provides a blast likened to that of a standard AM transistor radio turned on a local rock station with full volume. Whatever you do, however, make certain that both R1 and R2 are concealed *within* the box.

If you do not mind paying the postage and taking a bit of risk that you will never see this machine again, it makes an ideal gift. Just turn it on before you mail it. Battery drain is relatively low, and if you parallel several 9-volt transistor radio batteries, it will scream, wail, and blink lights for months without any ill effects— except to your victim. See Fig. 36 and Table 36.

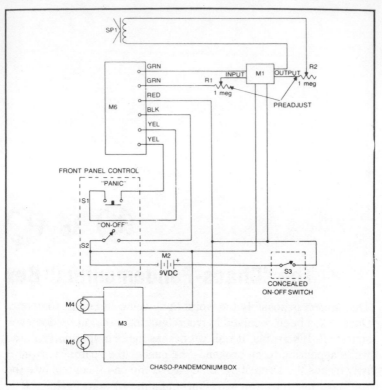

Fig. 36. Chaos-pandemonium box.

Table 36. Parts List for Chaos-Pandemonium Box.

Item No.	Description
M1	Three-transistor amplifier. (Lafayette 99 C 9039 or equiv.)
M2	9-VDC battery.
M3	Dual-flasher module. (Lafayette 19 C 0106 or equiv.)
M4, M5	No. 47 bulbs.
M6	Siren module. (Lafayette 19 C 0105 or equiv.)
R1, R2	1-megohm potentiometers.
S1	Normally-closed push button.
S2	Delayed-action switch. (Lafayette 34 R 3805 or equiv.)
S3	Subminiature spst switch.
SP1	Transistor radio speaker. (Lafayette 99 T 6032 or equiv.)

 37

Gypsy Light Organ

Here is a dandy little gadget that will enable the skilled to play music without even touching the "organ"! Working entirely on a light basis, the shadow of your hands moving over the two photocells produces weird musical effects, particularly if you space the two photocells in such a way that permits your using both hands, one photocell for each.

The heart of the gypsy light organ is the unijunction transistor oscillator module, available from moist major electronic parts suppliers. In conjunction with the two photocells shadows vary the amount of voltage reaching the module; hence the variable tones are produced.

You can feed the audio output into a standard hi-fi audio amplifier, or wire in a small transistor amplifier.

Anyone who has a good musical ear should be able to learn many song selections and numbers on this light organ. See Fig. 37 and Table 37.

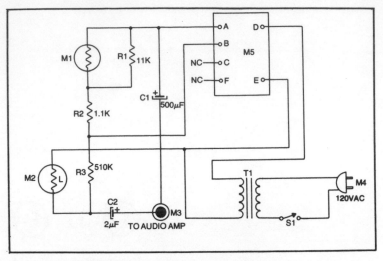

Fig. 37. Gypsy light organ.

Table 37. Parts List for Gypsy Light Organ.

Item No.	Description
C1	500-μF, 25-wVDC electrolytic capacitor.
C2	2-μF electrolytic capacitor.
M1, M2	Photocells. (GE Type X6 or equiv.).
M3	Microphone receptacle with locknut.
M4	AC wall plug, cap, and cable clamp.
M5	Unijunction transistor oscillator module.
	(Midland Standard Type 5100-4A.)
R1	11K resistor.
R2	1.1K resistor.
R3	510K resistor.
S1	Spst switch.
T1	Isolation transformer, 120:120 volts.
	(Allied 54 C 1426 or equiv.)

38

Hi-Fi Color Lights

Are you interested in adding unusual visual effects to your hi-fi, stereo, or FM radio radio system? This is possible with the gadget outlined in the accompanying diagram. By plugging in a lamp (using colored light bulbs, if you like, up to a total of 200 watts), the brightness of the light is actually controlled by the audio signal. Thus; as the lilting strings begin, the colored lamp is just barely glowing; when the bass drum is hit and the brass begins blaring, the corresponding bulb emits a more intense light. The dancing color lights cannot be fooled. If a sharp difference in sound occurs, the same will be noted visually the instant it occurs.

Construction is not difficult and the unit can be wired into a very small container, complete with RCA phono plug for audio input and a standard AC socket for 120 VAC output. Take care to avoid shorts; note that at no time is any portion of the circuit grounded to the chassis. See Fig. 38 and Table 38.

Testing can be accomplished with a 30- or 35-watt bulb plugged into the AC output socket. With no audio present, a slight glowing should be present. A low-level audio signal should drive the bulb to nearly full brilliance.

For best effect, you'll probably want this glowing to disappear when no audio is on the line. You can vary the "firing point" of the light bulbs by substituting a 100K potentiometer for R2; generally, the more resistance you insert, the less the resultant glow.

Although the circuit will drive a 200-watt load, you may find over-all operation cooler if it is kept down to, say, 125 watts.

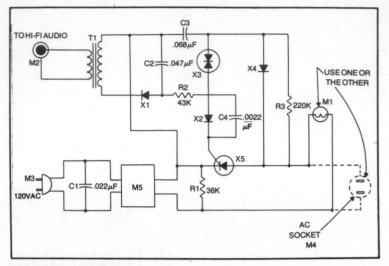

Fig. 38. Hi-fi color lights.

Table 38. Parts List for Hi-Fi Color Lights.

Item No.	Description
C1	.022-μF capacitor.
C2	.047-μF capacitor.
C3	.068-μF capacitor.
C4	.0022-μF capacitor.
M1	Colored household (120-VAC) bulb, to 150 watts.
M2	Microphone receptacle with locknut.
M3	AC wall plug, cap, and cable clamp.
M4	AC socket.
M5	Full-wave bridge rectifier module. (Erico FWB 3006A or equiv.)
R1	36K resistor.
R2	43K resistor.
R3	220K resistor.
T1	4:10K-ohm output transformer. (Thordarson TR-203 or equiv.)
X1, X2, X4	1N4003 silicon diodes.
X3	TI-42 trigger diode.
X5	2N3528 SCR.

90

39

Electronic Rainmaker

Ever find yourself *wishing* it were a rainy day just so you would be able to get to sleep easier? The circuit in the accompanying illustration will duplicate exactly the sound of a soft rainfall. It has been "lifted" from a device used by sound-effects men for use in taping radio shows where a rain sound must be injected. Reworked for personal application, the gadget is a two-tube table-model adapter for use with a tube-type AM or FM radio. The only use which the radio serves, actually, is a source for the required 150 and 250 VDC required to power the electronic rainmaker. The prime rainmaker unit contains its own speaker, and can be adjusted both for desired volume and for the kind of a rain you'd like to duplicate. With the circuit here, you can simulate anything from a light spring shower to a crashing, fierce summer thunderstorm.

Heart of the unit is a Solitron SD-1W white noise generator, although other types will work as well. The noise is injected as desired through R12, the "rain control." Note that the circuit contains extensive capacitive padding which accomplishes the softness required for true duplication of the genuine weather condition. See Fig. 39 and Table 39.

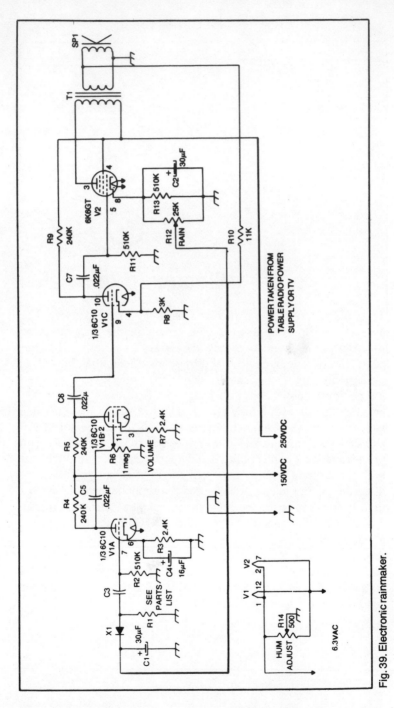

Fig. 39. Electronic rainmaker.

92

Table 39. Parts List for Electronic Rainmaker.

Item No.	Description
C1, C2	30-μF, 450-wVDC electrolytic capacitors.
C3, C5, C6, C7	.022-μF capacitors.
C4	16-μF, 450-wVDC electrolytic capacitor.
R1	Load resistor specified for specific diode employed, called out in diode instructional sheet.
R2, R11, R13	510K resistors.
R3, R7	2.4K resistors.
R4, R5, R9	240K resistors.
R6	1-megohm potentiometer.
R8	3K resistor.
R10	11K resistor.
R12	25K potentiometer.
R14	500-ohm potentiometer.
SP1	4-ohm speaker. (Utah SP25A or equiv.)
T1	7000:4-ohm output-transformer. (Stancor A-3878 or equiv.)
V1 (A,B,C)	6C10 compactron.
V2	6K6GT tube.
X1	White noise diode. (Solitron SD-1W or equiv.)
	One compactron tube socket. (Cinch Jones 12CS-B or equiv.)
	One standard 8-pin tube socket. (Amphenol 88-8X or equiv.)

 **40**

Do-Nothing Kiddie Box

Here is another of those toys that is guaranteed to amuse the youngsters for hours on end. Containing two GE-1 transistors and a garden-variety 2N105 plus a raft of assorted switches and potentiometers, the device will emit squealing sounds at various tonal levels through a speaker, flash lights alternately, simultaneously, or whatever, plus mystify the youngsters as to how to get these things to occur. Unless, for example, the correct "combination" is known, it will take a bit of experimentation to determine the correct settings of switches S1, S2, and S3 required to make the oscillator cease operation and the lights to flash, or both. To further add pandemonium, you should scatter the switches and potentiometer controls about the kiddie box, with at least one knob to turn on each side of the box. You can label the controls anything you like, but from experience let us suggest that you label them exactly the opposite of what they are supposed to do. For example, S2 can be labeled "sound off," which, of course, promptly sets the beeper going when closed. Use your imagination.

The author's recommendation is to build this into a sturdy wooden box, brightly painted. Secure all component installations to take a beating. Likewise check all solder joints for the same reason. Bulbs M3 and M4 can be mounted either on top of the box, or recessed in with transparent filters allowing color to come through. These are readily available under the Kodak label at any camera store, and they add a bit of brightness to the finished product. See Fig. 40 and Table 40.

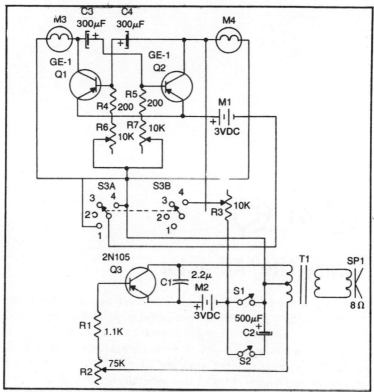

Fig. 40. Do-nothing kiddie box.

Table 40. Parts List for Do-Nothing Kiddie Box.

Item No.	Description
C1	2.2-μF, 35-w VDC capacitor.
C2	500-μF, 40-w VDC electrolytic capacitor:
C3, C4	300-μF, 40-w VDC electrolytic capacitors.
M1, M2	3-VDC batteries.
M3, M4	No. 49 bulbs.
Q1, Q2	GE-1 transistors.
Q3	2N105 transistor.
R1	1.1K resistor.
R2	75K potentiometer.
R3, R6, R7	10K potentiometers.
R4, R5	200-ohm resistors.
S1, S2	Spst switches.
S3 (A,B)	4-position, double-pole rotary switch.
SP1	8-ohm speaker. (Lafayette 99 T 6032 or equiv.)
T1	500:8-ohm output transformer. (Knight 54 C 2358 or equiv.)

 41

"Crazy-Lite" Attractor

Here is a great project for those electronic experimenters and others who enjoy psychedelic visual effects. Just plug any normal household lamp into our crazy-lite attractor and watch the action: It will flash repeatedly, then slow to a low glowing only to shut itself off and then sporadically get weak and intense. Periodically it will act in a normal fashion, although this generally occurs only for a few minutes after switch S1 is thrown on. This feature makes for interesting reactions when guests are in the living room. Properly wired permanently into a table lamp (appropriately attired with a bright red bulb), the antics will start soon after you have asked your guest to do you a favor by turning on that particular lamp. First, he'll have something to say about the red bulb, then he probably will not be saying very much at all for awhile as he watches in amazement your lamp do everything but take off and fly across the room. Requiring only one unijunction transistor, the attractor can be built into the base of most lamps.

If you desire added versatility, you can build it as shown in Fig. 41, complete with AC socket for plugging in the lights. Although it is not generally a good idea to plug in heavy appliances, you can have a great deal of fun with small devices such as portable vacuum cleaners, mixers and blenders, etc. Also, see Table 41.

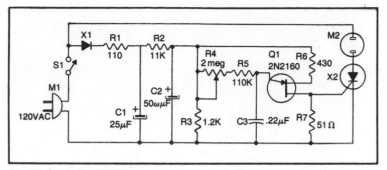

Fig. 41. Crazy-lite attractor.

Table 41. Parts List for Crazy-Lite Attractor.

Item No.	Description
C1	25-μF, 150-w VDC electrolytic capacitor.
C2	50-μF, 150-w VDC electrolytic capacitor.
C3	.22-μF, 75-volt molded capacitor.
M1	AC wall plug.
M2	AC socket.
Q1	2N2160 UJT.
R1	110-ohm, 1-watt resistor.
R2	11K, 10-watt resistor.
R3	1.2K, 12-watt resistor.
R4	2-megohm potentiometer.
R5	110K resistor.
R6	430-ohm resistor.
R7	51-ohm resistor.
S1	Spst switch.
X1	1N2069, or 200-PIV, 750-mA silicon diode.
X2	GE-X1, or 200-PIV, 5.0-A average-forward-current SCR.

 42

Match-Needing Electric Lamp

For more fun with electric lamps, build this novel gadget and watch the reaction you will get. After throwing the switch "on," there is virtually no way to get the bulb to light except by lighting a match and bringing it close to the concealed photocell, M3. Once the photocell senses the light, it causes bulb M2 to light, and the light from this bulb landing on the photocell *keeps* the lamp on until you open switch S1.

Properly built into a small standard desk lamp fixture or table lamp, you will be able to mystify friends every time you light up. Of course you will have to rewire the lamp to house the circuit, and rig it with a 6 or 1.25-volt bulb instead of 120-VAC type, but you'll still get a lot of light and have the added feature of automatic response to a lighted match.

The secret of doing this correctly is to locate the photocell in such a position that it will not pick up stray light from other sources, yet will respond to both the light from the bulb and your match. Trail and error works best in this case, as positioning will vary from one installation to another. See Fig. 42 and Table 42.

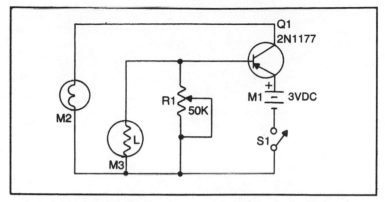

Fig. 42. Match-needing electric lamp.

Table 42. Parts List for Match-Needing Electric Lamp.

Item No.	Description
M1	3-VDC battery.
M2	1.25-volt bulb. (GE-123 or equiv.)
M3	Photocell. (RCA 7163 or equiv.)
Q1	2N1177 transistor.
R1	50K potentiometer.
S1	Spst switch.

43

Junior-Size Tesla Coil

Ever have an urge to man one of those fantastic machines that Boris Karloff is always fooling with in old Frankenstein movies? Well, probably 90 percent of these films make use of hissing, zapping, Tesla coils for visual effects. And you can build a scaled-down version of these monstrosities and still wind up with a wild device capable of raising your hair (literally!), sparking over whenever a metal object is brought close to the needle, or any one of numerous pranks this infernal device is famous for.

The principle is simply one of creating a fantastically-high voltage capable of doing tricks only high-voltage can do, while retaining the safety factor of extremely low current generation.

The only adjustment you'll have to make on this gadget is to tune C3 for maximum snapping and zapping. When the proper position is reached, you should see electric sparks pouring forth from the point of the sewing needle, particularly when "drawn off" by an item such as a screwdriver, etc.

Keep the length of heavy wire connecting the sewing needle with L1 as short and perfectly straight as possible. It should *not* be designed with the needle wire acting as a flexible cable for a shock rod regardless of how tempting such an idea might appear. The heavier and shorter that wire is, the more resultant zap. Have the needle pointing straight up for best effect.

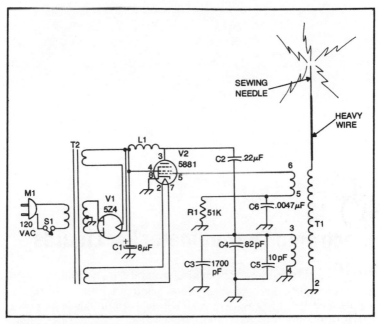

Fig. 43. Junior-size Tesla coil.

Table 43. Parts List for Junior-Size Tesla Coil.

Item No.	Description
C1	8-μF, 450-wVDC electrolytic capacitor.
C2	.22-μF, 600-volt molded capacitor.
C3	550-1700-pF timer capacitor.
C4	82-pF, 3-kV capacitor.
C5	10-pF, 3-kV capacitor.
C6	.0047-μF, 600-volt mica capacitor.
L1	30-mH rf choke. (J.W. Miller 692 or equiv.)
M1	AC wall plug.
R1	51K resistor.
S1	Spst switch.
T1	High-voltage transformer. (J.W. Miller 4526 or equiv.)
T2	Transformer. (Thordarson 24R09U or TV transformer with 120-VAC :500-VAC CT and 6.3-VAC filament windings.)
V1	5Z4 tube.
V2	5881 tube.

Electronic Thermometer For Liquids

Here is just the thing for making accurate liquid temperature readings, particularly for amateur photographers' darkrooms where it is difficult at best to read out degrees in Fahrenheit on a mercury-type thermometer. The gadget shown in Fig. 44 provides a close continuous check on the temperature of any liquid. Also, see Table 44. When the temperature varies either above or below where it should be, an alarm goes off in the form of a ticking relay. If you like, you can rig up an external buzzer or bell (by hooking it into the K3 relay contacts) to signal that something's gone amiss.

The probe that will be inserted into the liquid contains a thermistor, as shown in the circuit diagram. It can be housed in a hollow ballpoint pen, or whatever, so long as the thermistor wires do not become shorted underwater. The thermistor itself, however, should touch the liquid at some point.

The circuit is not really complicated or difficult to put together, although care should be taken during preliminary adjustment. This procedure should be accomplished as follows: Plug in the thermistor probe, set potentiometer R9 at minimum (to the left), and turn R2 fully open (to the right). Now turn the power on and gradually open R2 until you hear the relay clicking. At this point it has probably told you that your room temperature is on the order of 72 degrees Fahrenheit. Now advance R2 just a hair and ajust R9 (to the right) until the clicking stops.

Next thing you will want to do is paste a piece of paper behind the control knob (should be a pointer-type) of R2. You'll want to mark the calibration points with a pencil as you proceed.

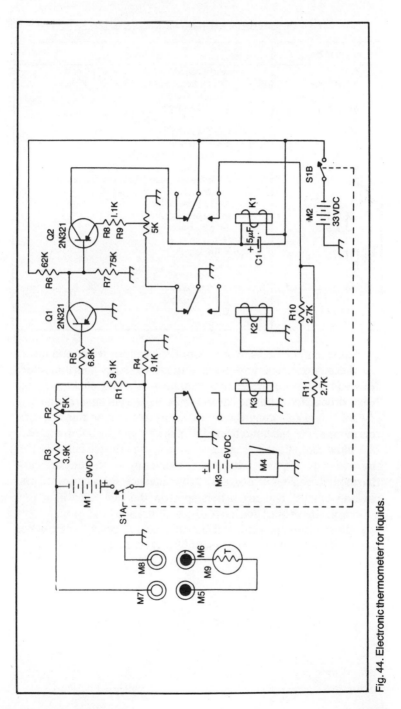

Fig. 44. Electronic thermometer for liquids.

103

Table 44. Parts List for Electronic Thermometer for Liquids.

Item No.	Description
C1	5-μF, 50-wVDC electrolytic capacitor.
K1, K2, K3	DC relays. (Sigma 11F-9000-G/SIL or equivalent.)
M1	9-VDC battery.
M2	33-VDC battery supply.
M3	6-VDC battery.
M4	6-volt buzzer.
M5, M6	Test plugs, single-prong.
M7, M8	Test jacks.
M9	Thermistor. (Allied Radio 60 C 8629 or equiv.)
Q1, Q2	2N321 transistors.
R1, R4	9.1K resistors.
R2, R9	5K potentiometers.
R3	3.9K resistor.
R5	6.8K resistor.
R6	62K resistor.
R7	75K resistor.
R8	1.1K resistor.
R10, R11	2.7K resistors.
S1 (A,B)	Dpst switch.

Now half fill a glass with room-temperature water and insert the thermistor probe and a conventional mercury thermometer. Now add a little hot water until you hear the relay clicking again. Next, drop an ice cube and advance R2 until the clicking resumes.

At this point, copy the temperature reading on the mercury thermometer on the homebrew dial-plate behind the pointer knob.

Now continue this procedure by dropping the ice cube back in the water to a point where the temperature on the mercury thermometer drops 2 degrees. Now advance R2 again until the clicking begins, and mark this spot on the dial. Continue this cooling business until you have all the calibrations you want.

The finished product will be calibrated to react to any water temperature you desire from about 58 to 86 degrees Fahrenheit.

45

Portable Metronome

Here is a dandy transistor metronome which you can build into a case as small as a tiny transistor radio, yet has all the features of the professional models you frequently see on top of pianos. In fact, an old nonworking transistor radio is the perfect housing for this gadget. In most cases, you can merely remove the main circuitry of the radio but retain the following parts: the on-off volume control potentiometer, speaker, battery and battery holder, and speaker transformer. If you have done this, you have just eliminated the following items in the accompanying schematic: 6/4 ohm speaker, S1, R1, M1, and T1. This leaves you with only four parts required to build the portable metronome—the 8-μF capacitors and 25-μF capacitors, the 68K resistor, and the 2N105 junction transistor. See Fig. 45 and Table 45.

Since costs were cut by eliminating the second transistor (which would have acted as an audio amplifier), the portable metronome runs wide open in normal operation. As soon as S1 is turned on, the metronome starts clacking. Control R1 determines the "beat" or frequency of clacking to which you wish to set the device.

To calibrate, merely synchronize your metronome with a commercial type or with any known source—including a phonograph record. Once you have jotted down the main beats such as "4/4" for four-four time, you're in business! If you're reworking a transistor radio using its volume control for the clack-rate control, you'll have to approximate settings or scratch markings on the side of the radio case along with a "key" on the knob.

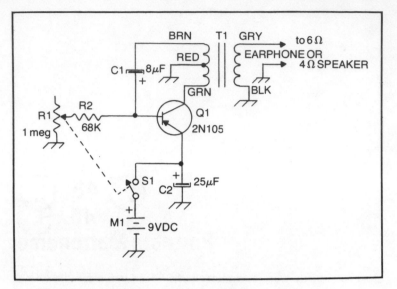

Fig. 45. Portable metronome.

Table 45. Parts List for Portable Metronome.

Item No.	Description
C1	8-μF, 25-w VDC electrolytic capacitor.
C2	25-μF, 25-wVDC electrolytic capaictor.
M1	9-VDC battery.
Q1	2N105 transistor.
R1	1-megohm potentiometer with spst switch.
R2	68K resistor.
S1	Spst switch on R1.
T1	Transistor radio output transformer. (La-fayette 99H 6123 or equiv.)

Personal Electronic Dripolator

Here is a gadget specially designed to convert sound sleepers to incurable insomniacs: a "drip, drip, drip" sounder guaranteed to drive your victim to distraction. Sort of a combination electronic metronome and transistor percolator, the circuit will power any convenient 3.2-ohm permanent-magnet speaker with sufficient volume to accomplish your purpose.

When S1 is flicked on, the dripping begins. You can adjust for most effective drip-rate by setting R2 to the desired point. Other variations can be achieved by trying substitutions for C1.

If a volume control is desired, add a 5K potentiometer between the junction of C1-Q1-R1 and SP1-L1. See Fig. 46 and Table 46.

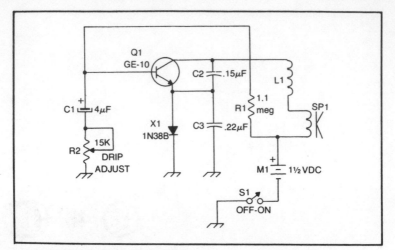

Fig. 46. Personal electronic dripolator.

Table 46. Parts List for Personal Electronic Dripolator.

Item No.	Description
C1	4-μF, 25-wVDC electrolytic capacitor.
C2	.15μF capacitor.
C3	.22-μF capacitor.
L1	100-mH miniature choke. (Lafayette Radio 34 H 8825 or equiv.)
Q1	GE-10 transistor.
R1	1.1-megohm resistor.
R2	15K potentiometer.
S1	Spst switch.
SP1	Any 3.2-ohm permanent-magnet speaker.
X1	1N38B diode.

47

Electronic Siren

Have you ever wondered how those police and emergency vehicles are able to create such an ear-splitting attention-getter-sound as they do with their sirens? They are really quite elaborate (and expensive), but if you would be willing to settle for a "manually-operated" version, you can construct this one.

As S2 is slid into action, the wailing begins. A low-frequency tone starts a shrill climb. (Incidentally, you can exercise some control over this by substituting a 75K pot for R2). For the true "siren" effect, merely depress S1 and watch the action. The climbing scream will suddenly reverse itself and start downward. To repeat the procedure, depress S1 again. After a few minutes' practice, you will have no problem achieving true simulation of the real thing.

If you like, you can install a volume control across the speaker leads. A 5K potentiometer should do the job nicely. See Fig. 47 and Table 47.

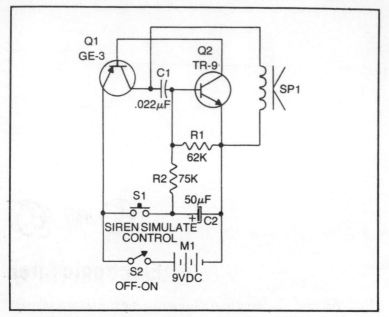

Fig. 47. Electronic siren.

Table 47. Parts List for Electronic Siren.

Item No.	Description
C1	.022-μF capacitor.
C2	50-μF, 15-w VDC electrolytic capacitor.
M1	9-VDC battery.
Q1	GE-3 transistor.
Q2	TR-9 transistor.
R1	62K resistor.
R2	75K resistor.
S1	Push-button spst switch.
S2	Standard spst slide switch.
SP1	Any 8-ohm permanent-magnet speaker.

110

⊕ 48 ⊕

Telephone Buzzer Toy

At last here is a use for those surplus telephone dial mechanisms just about everyone seems to have laying around gathering dust these days. This clever little toy will provide hours of pleasure for all groups of youngsters, although it is perhaps most amusing to children of ages two through five. The circuit contains no central on-off switch, so it is always ready for action. It is impossible for the device to inadvertently be left on, draining the battery, since the telephone mechanism always returns on an "off" position by itself. For these reasons, you should seldom have to replace M1.

Use a standard 6-volt buzzer, available through most parts stores or your local hardware shop. The bulb can be a regular No. 49 type (or whatever) 6.3-volt filament lamp or a 6-volt auto dash bulb.

Our suggestion is to build the buzzer toy into a sturdy wooden box, brightly painted. Remember it's going to get knocked around quite a bit, so make certain your batteries are securely locked into position.

Incidentally, close inspection of Fig. 48 will reveal that a jumper wire has been placed across the two upper terminals on the dial mechanism. Make sure this wire is secured in place. Also, see Table 48.

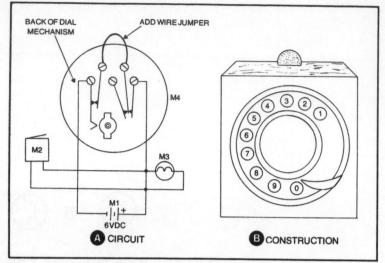

Fig. 48. Telephone buzzer toy.

Table 48. Parts List for Telephone Buzzer Toy.

Item No.	Description
M1	6-VDC dry cells.
M2	6-volt buzzer.
M3	6-volt pilot bulb.
M4	Telephone dial mechanism.

49

VLF Converter

Have you seen those gadgets which when placed next to your AM radio convert it for eavesdropping on specific vhf channels? Well, here's one that goes the other route—ultralow 200–400-kHz eavesdropping—and it does not require your buying a crystal for a specific frequency. Instead, you merely present it by tuning L2 to any VLF (very low frequency) signal. Next, place it next to your AM radio, and presto!

The construction is not critical, although the recommended coils should be installed and wiring made exactly as shown in Fig. 49. Remove the aluminum can surrounding T1 and position the coil similarly to however you elect to mount L2, although the two coils should be separated.

For tuneup, merely tape for external inductance coil (L3) to the case of your transistor radio as close as possible to the AM radio self-contained antenna coil. Now adjust the slug of L3 unit it is about ¾ inch outside of its coil form. Turn the power switch (S1) on and adjust L2 until you hear a signal in your AM radio on an unused frequency. Now adjust L3 and T1 for maximum reception.

You will find that you can "tune" the 200–400-kHz area by simply tuning the AM radio in a normal fashion. If a broadcast signal is encountered, readjust L2. Conversely, if you find a good blank spot on the AM radio and wish to stay there, tune L2 to the desired frequency in the 200–400-kHz band and then repeak T1 and L3 for best overall performance at that frequency.

Incidentally, if you build your converter in a wooden or plastic case, you will not have to make inductance coil L3 an external one. Just move the converter next to the AM radio and position it for best results.

A good external antenna and ground are a must for optimum results. String a long wire between a few trees and you will be surprised at the stations you will be able to hear.

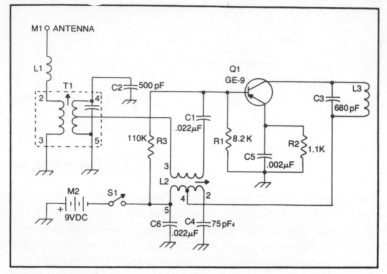

Fig. 49. VLF converter.

Table 49. Parts List for VLF Converter.

Item No.	Description
C1, C5, C6	.022-μF capacitors.
C2	500-pF capacitor.
C3	680-pF capacitor.
C4	75-pF capacitor.
L1	1-mH rf choke. (National R-50 or equiv.)
L2	455-kHz oscillator coil. (Miller 2020 or equiv.)
L3	Vari-loopstick. (Meissner 14-9015 or equiv.)
M1	Binding post.
M2	9-VDC battery.
Q1	GE-9 transistor.
R1	8.2K resistor.
R2	1.1K resistor.
R3	110K resistor.
S1	Spst switch.
T1	455-kHz i-f transformer with built-in capacitor. (Miller 2041 or equiv.)

Appendix A
Substitution Guide

The following are standard substitutions for most of the major components (tubes, transistors, and diodes) used in this book. These substitutions have not been tried in the circuits and some slight variation in circuit operation is possible and should be expected. Some substitutions will probably improve overall operation. Where no substitute is shown, no practical one was found. For general transistor substitutions see Table A-1.

The transistors listed in Table A-2 can be substituted for most of those shown in Table A-1. In many cases it is only necessary to check polarity called for in the circuit to determine which universal transistor can be used in a given situation. Also, see Tables A-3 and A-4.

Table A-1. General Transistors.

Type	Substitute
2N105	GE-2, 2N107
2N109	2N217, 2N270, GE-2
2N188A	2N241A, 2N320, 2N321, 2N396A, N414B, 2N414C, GE-2
2N217	2N109, 2N466, GE-2
2N278	GE-4
2N280	2N1066, GE-2
2N297	GE-4, SK3009
2N319	2N320, GE-2
2N321	25A397, GE-2
2N331	2N396A, 2N414B, 26440A 26597, GE-2
2N397	2N1316, 2N1348, 2N1349, 2N1354, 2N1355, GE-1
2N415	2N396A, 2N415A, 2N416, 2N440, 2N440A, GE-1
2N517	2N1316, 2N1969, GE-1
2N461	2N382, 2N383, 2N652, 2N1350, GE-2
2N555	GE-3
2N1177	GE-9
TI-415	TI-416, TI-419

Table A-2. Universal Substitution Transistors.

Transistor	Applications
GE-1	Pnp. Applications: Conv., mixer, osc., rf amp., i-f amp.
GE-2	Pnp. Applications: Audio amp.
GE-3	Pnp. Applications: Audio power output.
GE-5	Npn. Applications: Conv., mixer, osc., rf amp., audio-frequency amp.
GE-7	Npn. applications: I-f amp.
GE-9	Npn. Applications: Conv., mixer, osc., rf amp., i-f amp.
GE-10	Npn. Applications: Audio amp., audio output.
HEP-2	Pnp. Applications: General purpose.
HEP-3	Pnp. Applications: General purpose.
TR-9	Npn. Applications: Audio amp.

Table A-3. Diodes.

Type	Substitute
1N38B	1N38, 1N38A
1N540	1N1695, 1N2070, 1N1763

Table A-4. Tubes.

Type	Substitute
5Z4	5Z5G, 5Z4GT, 5AR4, 5CG4, 5V4G, 5V4GA, 5V4, 6087
6AQ8	6DT8 with filament wiring changes
6BM5	6AQ5, 6AQ5A. Also 6DS5 and 6DL5, provided filament wiring changes are made
6BX4	6AV4 with filament wiring changes
6K6GT	6K6, 6K6G
5881	6L6C, 7581, 7581A

Table B-1. Resistor Color Codes.

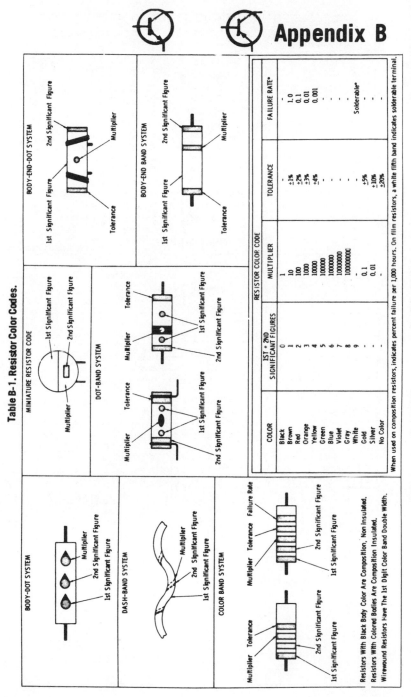

MINIATURE RESISTOR CODE — 1st Significant Figure, 2nd Significant Figure, Multiplier

BODY-END-DOT SYSTEM — 1st Significant Figure, 2nd Significant Figure, Multiplier, Tolerance

BODY-END BAND SYSTEM — 1st Significant Figure, 2nd Significant Figure, Multiplier, Tolerance

DOT-BAND SYSTEM — Multiplier, Tolerance, 1st Significant Figure, 2nd Significant Figure

BODY-DOT SYSTEM — Multiplier, 2nd Significant Figure, 1st Significant Figure

DASH-BAND SYSTEM — Multiplier, 2nd Significant Figure, 1st Significant Figure

COLOR BAND SYSTEM — Multiplier, Tolerance, Failure Rate, 2nd Significant Figure, 1st Significant Figure

Resistors With Black Body Color Are Composition, Non Insulated.
Resistors With Colored Bodies Are Composition Insulated.
Wirewound Resistors Have The 1st Digit Color Band Double Width.

RESISTOR COLOR CODE

COLOR	1ST + 2ND SIGNIFICANT FIGURES	MULTIPLIER	TOLERANCE	FAILURE RATE*
Black	0	1	-	-
Brown	1	10	±1%	1.0
Red	2	100	±2%	0.1
Orange	3	1000	±3%	0.01
Yellow	4	10000	±4%	0.001
Green	5	100000	-	-
Blue	6	1000000	-	-
Violet	7	10000000	-	-
Gray	8	100000000	-	-
White	9	-	-	-
Gold	-	0.1	±5%	Solderable*
Silver	-	0.01	±10%	-
No Color	-	-	±20%	-

When used on composition resistors, indicates percent failure per 1,000 hours. On film resistors, a white fifth band indicates solderable terminal.

Table B-2. Capacitor Color Codes.

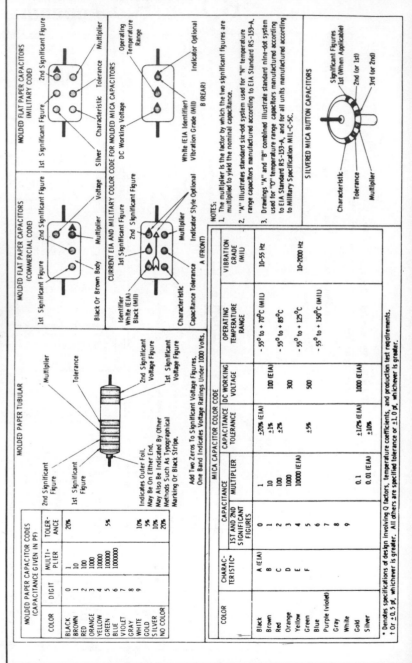

MOLDED FLAT PAPER CAPACITORS (MILITARY CODE)

1st Significant Figure, 2nd Significant Figure, Multiplier, Silver, Characteristic, Tolerance, Operating Temperature Range, White (EIA Identifier) Vibration Grade (Mil), Indicator Optional

A (FRONT) / B (REAR)

MOLDED FLAT PAPER CAPACITORS (COMMERCIAL CODE)

1st Significant Figure, 2nd Significant Figure, Voltage, Multiplier, Black Or Brown Body

CURRENT EIA AND MILITARY COLOR CODE FOR MOLDED MICA CAPACITORS

1st Significant Figure, 2nd Significant Figure, Multiplier, Indicator Style Optional, Characteristic, Capacitance Tolerance, DC Working Voltage, Identifier White (EIA) Black (Mil)

SILVERED MICA BUTTON CAPACITORS

Significant Figures 1st (When Applicable), 2nd (or 1st), 3rd (or 2nd), Characteristic, Tolerance, Multiplier

NOTES:

1. The multiplier is the factor by which the two significant figures are multiplied to yield the nominal capacitance.
2. "A" illustrates standard six-dot system used for "N" temperature range capacitors manufactured according to EIA Standard RS-153-A.
3. Drawings "A" and "B" combined illustrate standard nine-dot system used for "0" temperature range capacitors manufactured according to EIA Standard RS-153-A, and for all units manufactured according to Military Specification MIL-C-5C.

MOLDED PAPER TUBULAR

Multiplier, Tolerance, 2nd Significant Figure, 1st Significant Figure, 2nd Significant Voltage Figure, 1st Significant Voltage Figure, Indicates Outer Foil. May Be On Either End. May Also Be Indicated By Other Methods Such As Typographical Marking Or Black Stripe.

Add Two Zeros To Significant Voltage Figures. One Band Indicates Voltage Ratings Under 1000 Volts.

MOLDED PAPER CAPACITOR CODES (CAPACITANCE GIVEN IN PF)

COLOR	DIGIT	MULTI-PLIER	TOLER-ANCE
BLACK	0	1	20%
BROWN	1	10	
RED	2	100	
ORANGE	3	1000	
YELLOW	4	10000	
GREEN	5	100000	5%
BLUE	6	1000000	
VIOLET	7		
GRAY	8		
WHITE	9		10%
GOLD			5%
SILVER			10%
NO COLOR			20%

MICA CAPACITOR COLOR CODE

COLOR	CHARAC-TERISTIC*	CAPACITANCE 1ST AND 2ND SIGNIFICANT FIGURES	CAPACITANCE MULTIPLIER	CAPACITANCE TOLERANCE	DC WORKING VOLTAGE	OPERATING TEMPERATURE RANGE	VIBRATION GRADE (MIL)
Black	A (EIA)	0	1	±20% (EIA)	100 (EIA)	-55° to +70°C (MIL)	10-55 Hz
Brown	B	1	10	±1%			
Red	C	2	100	±2%	300	-55° to +85°C	
Orange	D	3	1000				
Yellow	E	4	10000 (EIA)			-55° to +125°C	10-2000 Hz
Green	F	5		±5%	500		
Blue		6				-55° to +150°C (MIL)	
Purple (violet)		7					
Gray		8					
White		9					
Gold			0.1	±1/2% (EIA)	1000 (EIA)		
Silver			0.01 (EIA)	±10%			

* Denotes specifications of design involving Q factors, temperature coefficients, and production test requirements.
† Or ±0.5 pf, whichever is greater. All others are specified tolerance or ±1.0 pf, whichever is greater.

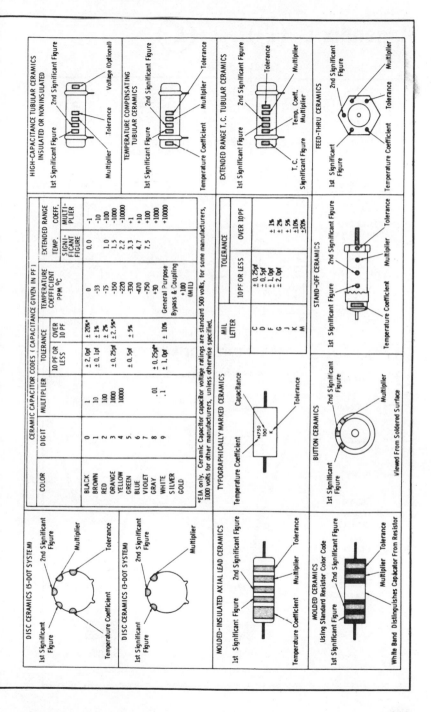

DISC CERAMICS (5-DOT SYSTEM)
1st Significant Figure · 2nd Significant Figure · Multiplier · Tolerance · Temperature Coefficient

DISC CERAMICS (3-DOT SYSTEM)
1st Significant Figure · 2nd Significant Figure · Multiplier

MOLDED-INSULATED AXIAL LEAD CERAMICS
1st Significant Figure · 2nd Significant Figure · Multiplier · Tolerance · Temperature Coefficient

MOLDED CERAMICS
Using Standard Resistor Color Code
1st Significant Figure · 2nd Significant Figure · Multiplier · Tolerance · White Band Distinguishes Capacitor From Resistor

HIGH-CAPACITANCE TUBULAR CERAMICS INSULATED OR NONINSULATED
1st Significant Figure · 2nd Significant Figure · Multiplier · Tolerance · Voltage (Optional)

TEMPERATURE COMPENSATING TUBULAR CERAMICS
1st Significant Figure · 2nd Significant Figure · Multiplier · Tolerance · Temperature Coefficient

EXTENDED RANGE T. C. TUBULAR CERAMICS
1st Significant Figure · 2nd Significant Figure · Tolerance · Multiplier · Temp. Coeff. Multiplier · T.C. Significant Figure

FEED-THRU CERAMICS
1st Significant Figure · 2nd Significant Figure · Tolerance · Multiplier · Temperature Coefficient

TYPOGRAPHICALLY MARKED CERAMICS
Temperature Coefficient · Capacitance · Tolerance
(N750 / 100 / K)

BUTTON CERAMICS
1st Significant Figure · 2nd Significant Figure · Multiplier · Temperature Coefficient · Viewed From Soldered Surface

STAND-OFF CERAMICS
1st Significant Figure · 2nd Significant Figure · Multiplier · Tolerance · Temperature Coefficient

CERAMIC CAPACITOR CODES (CAPACITANCE GIVEN IN PF)

COLOR	DIGIT	MULTIPLIER	TEMPERATURE COEFFICIENT PPM °C	EXTENDED RANGE TEMP. SIGNIFICANT FIGURE	EXTENDED RANGE COEFF. MULTIPLIER	TOLERANCE 10 PF OR LESS	TOLERANCE OVER 10 PF
BLACK	0	1	0	0.0	-1	±2.0pf	±20%*
BROWN	1	10	-33	1.0	-10	±0.1pf	±1%
RED	2	100	-75	1.5	-100		±2%
ORANGE	3	1000	-150	2.2	-1000	±0.25pf	±2.5%*
YELLOW	4	10000	-220	3.3	-10000		
GREEN	5		-330	4.7	+1	±0.5pf	±5%
BLUE	6		-470	7.5	+10		
VIOLET	7		-750		+100		
GRAY	8	.01	+30		+1000	±0.25pf*	
WHITE	9	.1	General Purpose Bypass & Coupling +100 (MIL)		+10000	±1.0pf	±10%
SILVER							
GOLD							

*EIA only. Ceramic Capacitor capacitor voltage ratings are standard 500 volts, for some manufacturers, 1000 volts for other manufacturers, unless otherwise specified.

TOLERANCE

MIL LETTER	10 PF OR LESS	OVER 10 PF
C	±0.25pf	
D	±0.5pf	
F	±1.0pf	±1%
G	±2.0pf	±2%
J		±5%
K		±10%
M		±20%

Appendix C

Table C-1. Tubes.

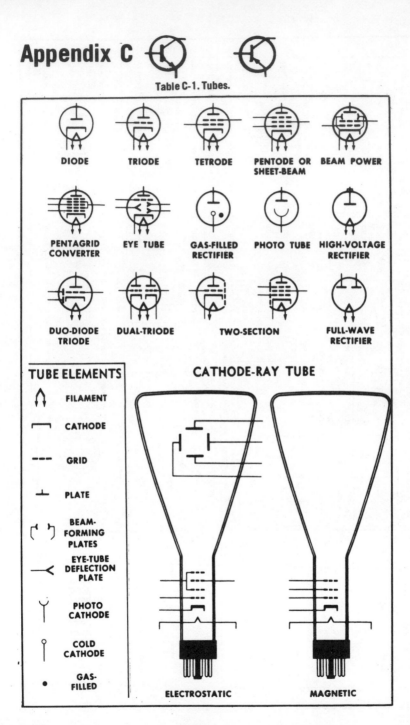

DIODE   TRIODE   TETRODE   PENTODE OR SHEET-BEAM   BEAM POWER

PENTAGRID CONVERTER   EYE TUBE   GAS-FILLED RECTIFIER   PHOTO TUBE   HIGH-VOLTAGE RECTIFIER

DUO-DIODE TRIODE   DUAL-TRIODE   TWO-SECTION   FULL-WAVE RECTIFIER

TUBE ELEMENTS

FILAMENT
CATHODE
GRID
PLATE
BEAM-FORMING PLATES
EYE-TUBE DEFLECTION PLATE
PHOTO CATHODE
COLD CATHODE
GAS-FILLED

CATHODE-RAY TUBE

ELECTROSTATIC   MAGNETIC

Table C-2. Semiconductors.

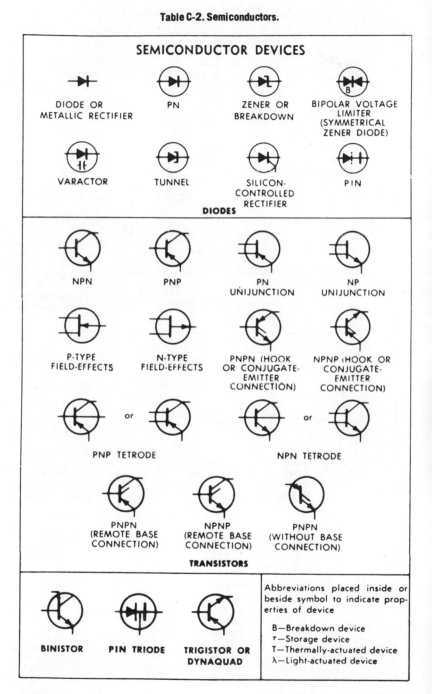

SEMICONDUCTOR DEVICES

DIODE OR METALLIC RECTIFIER

PN

ZENER OR BREAKDOWN

BIPOLAR VOLTAGE LIMITER (SYMMETRICAL ZENER DIODE)

VARACTOR

TUNNEL

SILICON-CONTROLLED RECTIFIER

PIN

DIODES

NPN

PNP

PN UNIJUNCTION

NP UNIJUNCTION

P-TYPE FIELD-EFFECTS

N-TYPE FIELD-EFFECTS

PNPN (HOOK OR CONJUGATE-EMITTER CONNECTION)

NPNP (HOOK OR CONJUGATE-EMITTER CONNECTION)

or

PNP TETRODE

or

NPN TETRODE

PNPN (REMOTE BASE CONNECTION)

NPNP (REMOTE BASE CONNECTION)

PNPN (WITHOUT BASE CONNECTION)

TRANSISTORS

BINISTOR

PIN TRIODE

TRIGISTOR OR DYNAQUAD

Abbreviations placed inside or beside symbol to indicate properties of device

B—Breakdown device
τ—Storage device
T—Thermally-actuated device
λ—Light-actuated device

121

Table C-3. Miscellaneous Symbols.

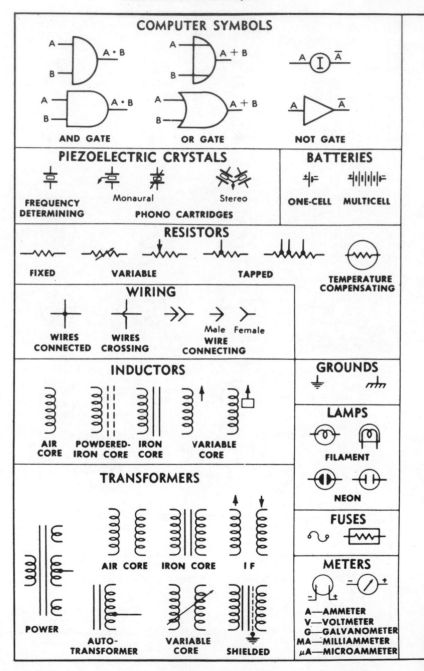

COMPUTER SYMBOLS

AND GATE OR GATE NOT GATE

PIEZOELECTRIC CRYSTALS

FREQUENCY DETERMINING

Monaural Stereo

PHONO CARTRIDGES

BATTERIES

ONE-CELL MULTICELL

RESISTORS

FIXED VARIABLE TAPPED TEMPERATURE COMPENSATING

WIRING

WIRES CONNECTED WIRES CROSSING Male Female WIRE CONNECTING

INDUCTORS

AIR CORE POWDERED-IRON CORE IRON CORE VARIABLE CORE

GROUNDS

LAMPS

FILAMENT

NEON

TRANSFORMERS

AIR CORE IRON CORE IF

POWER AUTO-TRANSFORMER VARIABLE CORE SHIELDED

FUSES

METERS

A—AMMETER
V—VOLTMETER
G—GALVANOMETER
MA—MILLIAMMETER
μA—MICROAMMETER

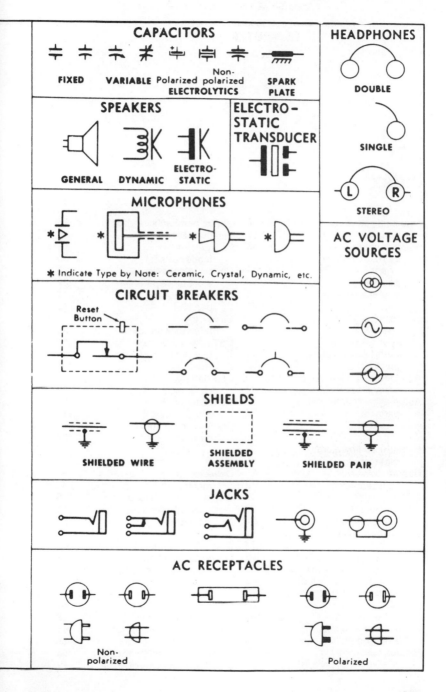

CAPACITORS

FIXED   VARIABLE   Polarized   Non-polarized   SPARK PLATE

ELECTROLYTICS

SPEAKERS

GENERAL   DYNAMIC   ELECTRO-STATIC

ELECTRO-STATIC TRANSDUCER

MICROPHONES

* Indicate Type by Note: Ceramic, Crystal, Dynamic, etc.

CIRCUIT BREAKERS

Reset Button

SHIELDS

SHIELDED WIRE   SHIELDED ASSEMBLY   SHIELDED PAIR

JACKS

AC RECEPTACLES

Non-polarized   Polarized

HEADPHONES

DOUBLE

SINGLE

L   R

STEREO

AC VOLTAGE SOURCES

Index

Edited by Roland Phelps